SCIENCE AND REVOLUTION

On the Importance of Science and the Application of Science to Society, the New Synthesis of Communism and the Leadership of Bob Avakian

An Interview with Ardea Skybreak

Published in 2015 by Insight Press, Chicago, Illinois.
Printed in U.S.A.

FIRST EDITION

ISBN: 978-0-9760236-1-6

Publisher's Note

In the early part of 2015, over a number of days, *Revolution* conducted a wide-ranging interview with Ardea Skybreak. A scientist with professional training in ecology and evolutionary biology, and an advocate of the new synthesis of communism brought forward by Bob Avakian, Skybreak is the author of, among other works, *The Science of Evolution and the Myth of Creationism: Knowing What's Real and Why It Matters*, and *Of Primeval Steps and Future Leaps: An Essay on the Emergence of Human Beings, the Source of Women's Oppression, and the Road to Emancipation*. This interview was first published online at www.revcom.us.

A Table of Contents can be found at the back of this volume.

SCIENCE AND REVOLUTION

On the Importance of Science and the Application of Science to Society, the New Synthesis of Communism and the Leadership of Bob Avakian

An Interview with Ardea Skybreak

A Scientific Approach to Society, and Changing the World

Question: I thought we would start by briefly asking some questions about science and the scientific method. So I actually wanted to start with kind of a provocative question: What does science have to do with understanding and changing the world? And, just quickly for some background on that, I think most people, including most natural scientists, don't think that you can, that you need to, or that you should take a scientific approach to analyzing society, or analyzing the "social world," much less changing it. So I wanted to ask you: Why is that notion wrong, what does science and the scientific method have to do with understanding and changing society and the world?

AS: Well, I think that's a very important question because, as you say, even many people who are scientists in the natural sciences and who apply very rigorous scientific methods when trying to deal with the natural world (biology, astronomy, physics, and so on), when you talk to them about **society**—the problems of society, the way societies are organized—all of a sudden it seems like their grasp of scientific method goes completely out the window! Many natural scientists actually start to revert then to a kind of crass **populism**, to just kind of talking vaguely about the "will of the people," or about elections, or some other things that really have little or nothing to do with analyzing in a scientific way the main features of a given society—how it's set up, how it functions—or with analyzing in a scientific way what's wrong in a society, or how societal problems could be solved in a scientific way. Not everyone is like that, but it's striking—the degree to which many advanced thinkers in the natural sciences seem to forget or drop everything they know about scientific methods whenever they try to think about the problems of society!

I think it's very, very important to understand that science as a method has not been around in the history of humanity for all that long. So people generally are simply not accustomed to trying to understand and transform reality in a scientific way. For most of the history of human beings on this planet, the understanding of both the natural and social world was derived more from a sort of basic trial-and-error approach, trying to figure things out catch-as-catch-can, and trying to solve problems that way—often making up all sorts of mystical and supernatural explanations to fill in the gaps in people's understanding. So, you know, people used to think lightning was the anger of the gods, or something like that, because for a long time they didn't have a scientific understanding of what actually caused lightning.

So I think it might be worth starting a little bit by talking about **what is science**, to demystify it a little bit. I mean, science deals with material reality, and you could say that all of nature and all of human society is the province of science, science can deal with **all** that. It's a tool—science—a very powerful tool. **It's a method and approach for being able to tell what's true, what corresponds to reality as it really is.** In that sense, science is very different than religion or mysticism, or things like that, which try to explain reality by invoking imaginary forces and which provide no actual evidence for any of their analyses. By contrast, science requires proof. It requires evidence. It is an evidence-based process. That's very important. **Science is an evidence-based process.** So whether you're just trying to understand something in the world, or trying to figure out how to change reality—for instance, you might be trying to cure a disease, or you might be trying to understand the dynamics of a rain forest or a coral reef ecosystem, or you might be trying to make a revolution to emancipate humanity, you know, the full range of material experience—science allows you to figure out what's really going on and how it can change.

I read somewhere that Neil deGrasse Tyson, in popularizing the importance of science, said something like—I'm paraphrasing here, but he said something like: Science allows you to confront and identify problems, to recognize problems and figure out how to solve them, **rather than run away from them**. And I think that's an important point, too. Science is what allows you to actually deal with material reality the way it really is. Whether you're talking about the material reality of a disease, of a natural ecosystem, or of a social system that human beings live under, science allows you to analyze its components, its history, how it came to be the way it is, what it's made of, what are its defining characteristics and underlying contradictoriness (and we'll come back to that) and therefore also what is the basis for it to change, or to be changed, if your intent is to change it. Whether you want to cure a disease or make a better society, you need that scientific evidence-based process.

One thing about science is that **it asks a lot of questions** about how things came to be the way they are, and about how things have changed over time. I've always been very interested in what's called the historical sciences—for instance, biological evolution, but there are also other historical sciences, including the science of human society—which deal with how things change over time. And then, of course, if you're studying how things change over time, you can study how things can change some more, including in directions that human beings might be inclined to have it go. All of reality has evolved, has changed over time, and it's still changing all of the time, whether you're talking about the natural world or the social world. If you want to change life, if you want to change the way a society is organized, if you want to change the world, if you want to change anything in nature or society, you need a scientific method, because that's the only way to deeply and systematically uncover how reality really is, on the basis of systematic observations and interactions, manipulations, and transformations of reality. That's how you learn how things really are, how they got to be that way, and how they can be changed. Again, it's an evidence-based process, it's not just "what you think" or "what I think." We need evidence, accumulated over time. This is what reveals what reality is made up of, how it came to be a certain way, how it may be changing right now, or how it may be possible for human beings to further change it.

Here's an important point: Without science, you can only say what you as an individual think reality is, or maybe you can say what a whole bunch of people think reality is, or maybe you can say what a government, or religious authority, or some other authority might tell you reality is like, but that doesn't make any of it true. **Without science you are at the mercy of being manipulated, of having your thinking manipulated and not being able to tell what's right from what's wrong, what's true from what's false.** If you really want to know what's what, what's true, and what to do, you **need** science—not fantasies or wishful thinking, but concrete evidence and a systematic process, a systematic method of analysis and synthesis. The analysis breaks down experience and knowledge over time; synthesis brings it back together in a higher way, in a more systematic way, getting the bigger lessons, the core lessons out of the accumulated experience.

So this is one of the reasons why you need scientific revolutionary theory if you really want to change a society at its roots. You know, we talk about radical change in society. Well, the word "radical" comes from the Latin meaning "root"; it means get to the root of the problem. Don't just stay on the surface of what the problem appears to be, on a superficial level or at just one moment in time. Get underneath it, get deeper, the way a good scientist does, to understand what are the deeper rules of the system, what are the deeper ways the contradictions inside a system make it work certain ways that cause problems, or that can bring forth possibilities.

Question: Well, if I could interject just for a second, this strikes me as really important and critical in terms of what is science and what's involved in a scientific approach to reality; what you're saying about the importance of science being evidence-based and the different points you were making about that, I think are very important there. One thing I wanted to interject is to kind of zero in on this question: I think a lot of people would recognize, including a lot of natural scientists—and obviously you, yourself, were trained as a natural scientist, and so maybe you would have some particular insights on this—but a lot of even natural scientists would probably look at what you were saying and respond, OK, I see how that process can be applied to the natural world, to the natural sciences—patterns, looking for evidence, synthesis—but then they would kind of recoil at the idea that you could actually apply that to human beings and human society. Or maybe another way to go at it is that some people would say, Well, OK, but human beings and human societies, that's just too complicated to be scientific about or to apply science. So maybe we could zero in a little bit on what does it mean specifically to take a scientific approach to human beings and human society and their development, and why is that correct?

AS: Well, look, for one thing, in any system, whether it's in the natural world or human society, there's both complexity and simplicity. The idea that human beings or human societies are just too complex to analyze with science is ridiculous. It's the exact opposite. How could you possibly deal with the complexity of human social organizations and interactions over various historical periods and up to today, and all the contradictions within that, all the complicated patterns and things, and the different forces, and so on, and different objectives of different peoples and different periods of history—how could you deal with all that without science? How could you even begin to make sense of it and understand it? And it's not true that natural systems are somehow simpler, you know. If you want to understand the dynamics of complex ecosystems—like, for instance, a rain forest, which has many different layers of trees and shrubs in the undergrowth and so on, and which is characterized by very complex dynamics in terms of the many different kinds and levels of interactions among and between the incredibly diverse plant and animal species—I mean, you could spend a lifetime, and many people do, just trying to get a beginning understanding of a lot of these complex dynamics. Or, if you wanted to better understand coral reef ecosystems, or desert ecosystems, or the differences between different ecosystems and which ones might be more vulnerable to being disrupted and which ones might be relatively more stable, or assess relative species diversity or how to preserve diversity... so many questions worth exploring further.... Look, I'm not trying to get into all that right now because I know you want to talk mainly about human social systems, but what I am saying is that in both the natural and social world, material reality is very complex, and that while we as human beings always have some shortcomings in our understanding (things that at any given time we don't quite get yet) we **also** have tremendous abilities and a lot of accumulated knowledge. Our brains are capable of actually investigating and exploring all sorts of questions, from many different angles, and we're actually capable of summing things up over a period of time, accumulating historical experience and knowledge that way. This is one of the things that's very particular to human beings: our great ability to accumulate understanding over generations, over centuries, over millennia, and to understand some of the patterns of organization of societies or of natural systems or whatever we turn our minds to.

And we humans are also capable of doing some very important projections into the future, not just the future tomorrow, or of a month from now, but also trying to understand what could be happening to this planet, for instance—the entire planet—from an environmental standpoint, looking ahead generations, not just tomorrow. Similarly with social systems, we actually have the ability to analyze different patterns of social organization throughout past human history and up through today, and we can also project ahead to the way things could be in the future. We can therefore also make some conscious decisions about what we want to work on now—in which direction do we want to try to push things, because we do have conscious initiative to do that. So, for instance, when you talk about a human society, about human social organization, you can see that a society is basically a way that human beings come together—work together, or oppose each other or whatever—but come together to essentially work on meeting the requirements of life of people in a given time. It might be done well, or it might be done poorly, but this is what a human society is, it's a form of organization. Right? And, you know, we've all lived in this capitalist-imperialist world for so long, those of us who are alive today, that sometimes it's hard to remember or to think about the fact that human societies haven't always been organized this way, and they don't have to be organized this way.

Capitalism-imperialism is not the only way to organize a human society, and I would argue strenuously that it's certainly not the best way. But in any case it's not the only way, and that is worth understanding and thinking about. The fact is we can apply science to try to understand some of those earlier social systems. For instance, many societies in the history of human beings were organized on the basis of **slavery**, the exploitation of slaves, the domination of slaves who were literally the property of the slave-masters, and the slavemasters made them build the economy that way. And I won't get into all the details of it, but that's a very different kind of society than the ones that mainly prevail today, on a large scale at least. There's still slavery in the world, by the way, including sexual slavery, which is a very big problem. But the fundamental and dominant forms of organization of societies in the world today are mainly not organized on the basis of slavery. But for a long time in the history of human beings, that was a dominant form of social organization.

Another significant form of social organization was the system of **feudalism**, and there are certainly still remnants of feudalism in many parts of the world today, we see it everywhere. But in feudal systems you had lords and masters, you had nobilities, you had aristocracies, and you had oppressed and dominated people like serfs and peasants, who would typically be growing the crops and having to turn much of it over to the lords of a region or whatever, and they had to pay terrible taxes and tributes to the lords, and they were just barely one notch above being outright slaves. It was even very common for a serf to have to turn over his daughter to the local lord of the region, to basically have sex with and do with whatever he will and there was nothing serfs could do about any of that under the existing rules of the feudal system. Feudalism in turn is a very different system than what's called **bourgeois democracy**, the kind of more typical **capitalist-imperialist system** of social organization that dominates the world today. I'm not going to try to get into any of this in detail right now, but I will say that it is worth thinking about the fact that scientific methods can be—and have been—applied to analyzing the patterns of social organization of all those different past social systems; and if we can do it for the past, we can also do it for the future.

Some people will say, well, OK, systems such as slavery, the feudal system, and maybe even the capitalist system, are not good ways to organize society, but what we should really do is just go back to an early communal system. Such people argue that we just need to organize on a small scale in our local areas, so that people can work together in small groups, and make all the decisions together, and can create "genuine democracy" and make collective decisions about how to meet the needs of the people, and promote local agriculture, local production, and so on. The problem with such views is that they are simply not rooted in the actual reality of the world today! Look, I would agree that there's a lot we could still learn from hunter-gatherer societies that prevailed for most of the history of humanity, that there's a lot we could still learn from some remnants of those societies in the world today, and that there's a lot we can learn from people today who have all sorts of ideas about how better to organize things, in a more rational way, on a relatively small and local scale, in terms of such things as agricultural production, waste reduction, promoting use of local products, and so on. So yes, there are things that we can learn from some of the social experiments that people are doing, trying to figure out how to get away from some of the problems of modern society that cause natural and social dislocations, pollution, the destruction of soils, and so on and so forth. **But let's get real, OK? We need to talk about the scope and scale of the human species spread out throughout this entire planet. Billions and billions of people. You're not gonna resolve the problems of society by going backwards to some kind of idealized, romanticized primitive communalism!** So if that's not going to cut it, if that's not going to be able to meet the key and critical problems of today, and certainly not with sufficient scope and scale, then what? Look, a slave-based system, a feudal system, a capitalist-imperialist system, these are all just material ways of organizing human societies and they can all be analyzed by science and critically evaluated. But you can also apply the same scientific methods to figuring out how to build completely new and different societies that would not only be better, but also be able to **encompass the whole planet**. Because I'm really not interested in talking about philosophies and methods that cannot, ultimately, encompass and benefit **all** of humanity.

One of the things you get from Bob Avakian [BA] which I really appreciate is that he's promoted this concept that we need "emancipators of humanity" and that we need to move in the direction of making this world, this entire planet, a good place to live in and function for all of humanity, where we can get away from the idea that some groups of people, and some categories of people, or some whole countries, are lording it over others, and exploiting and dominating and oppressing others. That's the whole idea of this revolutionary communism, and one of the things you really get from BA is the need to always think and proceed back from the need to emancipate all of humanity. Otherwise, you can easily fall into things that go off track. BA has talked about how **the goal is not for the last to be first and the first to be last**, it can't be about revenge, about the oppressed taking revenge on people. I agree that's not the kind of world we should be striving for. And my point here is that without science you're going to be lost, because without a scientific method to analyze the patterns, to really understand **why** things are the way they are and **how** they could be different, and **on what basis** could they be different, you're going to go off track all the time.

You know, one of the hallmarks of good science—because there is bad science, too—but one of the hallmarks of good science is really having a critical spirit and promoting critical thinking—which, by the way, is another hallmark of BA's work. He's really stressing the need for everyone to get into this—it doesn't matter what your level of education is...I would like to talk about this. Science is not something that should only be done by an elite, or by people who have gone to graduate school or gotten Ph.D. degrees or something like that. I firmly believe—and I can provide evidence of this—that **people who are not even trained in basic literacy can actually function as scientists**. You know, you can train people in scientific methods, in even just a weekend you can start to do that. If you want to get people doing science in the natural world, you can spend a weekend doing some good science in a rain forest or in a desert, and I guarantee you it will be <u>real</u> scientific work, <u>real</u> scientific investigation. And I don't care if you don't even have a sixth-grade education. If you are a healthy human being, you can take up and apply scientific methods, whether to the problems of nature or of human society. And one of the things I'm very concerned about is that we promote scientific understanding and scientific methods very, very broadly, **so that everyone can learn to use these methods**, and it's not just the province of a few or a province of the elites.

A Scientific Outlook, A Boundless Curiosity About the World

Question: Well, you just touched on something I wanted to ask you about. Very frequently the way science is portrayed and is viewed—and I think this relates to the point that you were just making about science being portrayed as the province of the elite—it's also often portrayed as cold, boring, lifeless, dry, maybe even some people think of it as being dogmatic or rigid, or something a relatively small number of people are practicing, kind of cut off from the world. And so I wanted you to respond to that view and portrayal of science.

AS: Oh boy, don't get me started! [laughs] I mean, at the risk of sounding ridiculous, some of the most passionate and lively people I've ever known have been scientists, including in the natural sciences. Science itself is not...how can anybody think of it as being dry or lifeless, or whatever, when the whole point of science is to have boundless curiosity about the world, about everything, about the way things came to be. Where did we come from? Where did life on earth come from? How did it come together? Why is this bird building its nest in this way in this tree and what is it doing? And what is this cat doing running across the road? I'm not trying to get into a lot of questions right now, but the point is that a good scientist is constantly asking questions about everything. It's what is often so wonderful about little kids, how little kids want to know everything about everything: **why** is this like this, **why** is this like that, **how** did it come to be that way, what **is** this? And so on. And unfortunately that natural scientific curiosity that pretty much every kid has, often gets kind of sucked out them, beaten out of them—if not physically then just through the stultifying educational system, and through the way this society is, and what it encourages and discourages.

Why do so many people think of science as something scary or dry or lifeless? Frankly, it's for a number of reasons. One, they often haven't been taught correctly in schools what science **is**. Science is sometimes taught as if it's just a bunch of dry precepts or formulas—just a bunch of end-point conclusions people are supposed to remember—but that's not science. Science is a **process**. It is a lively **method of investigation**. Think of science as a way that allows you to ask a whole lot of questions, about everything and anything, and that gives you a method and approach to enable you to systematically and methodically investigate things, to act sort of like a detective out in the world, to deeply investigate natural reality, or social reality. There's nothing lifeless about it! It's all about trying to understand things, including because of the basic principle that if you want to change anything you'd better first understand it, and not understand it just in a superficial way. You need evidence, accumulated over time, and not just in scattered little bits and pieces. You need to discover **the patterns, including the patterns of how things relate to each other**: If you want to understand the interactions between, say, oak trees and the squirrels that disperse their acorns; or between some of the flowering plants and the bees or butterflies or birds or even monkeys that may act as their pollinators; or between sharks and their prey, just to use a few examples—if you want to understand any of this, you need to uncover the evidence of the underlying patterns and the underlying dynamics and you need science to do this. Life is full of dynamic interactions—not just in that broader natural world, but in the human social world as well. So if you want to change anything, you first really have to understand why things are the way they are, how they came to be that way, and which way they are moving. And if you don't like the way it's going, and it has to do with human society, **then do something about it**, by using human conscious influence to try to change the course or direction of things. That's what gets done whenever scientists come up with a cure for a disease, or figure out something like how a badly damaged river ecosystem might be reinvigorated by periodically releasing water from the dam that caused all the damage.

Well, these are some examples of science applied to the natural world, and I could give you millions of similar examples. Science is all about understanding the nature of things, understanding patterns, **and understanding transformation**—the way things get transformed even on their own, how things move, thanks to their internal dynamics and the effects of outside influences...you see, everything is always moving, material reality is always moving. Whether you're talking on the scale of the cosmos, the planets, the galaxies, or whether you're talking, on a more micro scale, about ants in an anthill or cells in your body or subatomic particles, **everything in material reality is always moving and changing, nothing ever stands still**. And when it comes to social life, human beings should be using the same methods of science to understand how societies got to be the way they are, and to analyze—scientifically—what's wrong with them; to analyze—scientifically—how could they be better; and to determine what would be a strategy for moving in the right direction—again, on a scientific basis.

Another reason people are sometimes turned off by science is because there has been **bad** science. There will always be "science" that's misused and misapplied, you know, but it's bad science, OK? For instance, take examples about the way sometimes in the course of history science has been used to promote the idea that some races are inferior to other races, are mentally inferior, or something like that. Well, that's **junk** science. In fact you can use rigorous scientific methods to prove that that was all bad science. It's not just "morally" bad—it is that, but it's also scientifically bad—it's completely false and **you can use good science to prove that**.

A Scientific Assessment: The World Today Is a Horror for the Majority of Humanity—And That *Can* Be Radically Changed

Question: Well, let's keep going on this point about applying science to understand why the world is the way it is and how it could be different, and what could be done about that. Looking at the state of the world right now, in two senses—one, in a more overall sense, in terms of what are the conditions that the vast majority of humanity find themselves in right now, what is the state of the world in a more overall sense, but then, kind of zeroing in on one particular dimension of that, obviously it's been very heartening these last few months that there have been things that we haven't seen in this society in the U.S. in quite a while, in terms of massive resistance to this epidemic of police murder and police brutality, concentrated in the murders of Michael Brown and Eric Garner and the grand jury decisions letting their killers go, with tens and tens of thousands of people directly in the streets around this, disrupting business as usual, and then millions of people here and around the world confronting all this—what I'm getting at is how would we apply science, both to the particularity of this moment and understanding that, but also looking in a more big picture sense at, as you were saying, why is the world this way and how it could be different?

AS: Well, I would start off by saying, OK, let's apply science to talking, first of all, about where humanity's at, what's the state of the world, what's the state of this society that we live in. And it's been said many times, including by BA, that the world, as it is, is a horror. Right? Now, this is being said by people, including BA, who are overall very appreciative of a lot of beauty in the world. Speaking for myself, trained as a biologist, as a natural scientist, I see beauty everywhere in the natural world, and among people, in the great diversity and richness of human experience and all the many different cultural expressions and the great variety of life, including social life. There is great beauty. But at the same time, it's undeniable: The world is a horror for the majority of humanity at this point in history.

Now, let's take the question of human suffering. It would be unscientific to think that you could ever completely eliminate human suffering. There will always be loss, there will always be death, there will always be grief, there will always be some forms of disease or some forms of catastrophes that negatively impact human beings. I don't think you could ever say you would get to the point where there would never be any human suffering; that would be a completely idealized false world and illusion. But what you can say, is that it is possible to get to a world that is not characterized by so much unnecessary suffering.

And the reality of the world today—I mean, look at this society, what you were just talking about, all these police murders. You know, I can't take it any more—and I won't take it any more! Practically every single day, you hear about another person, usually Black or Latino, male, unarmed, who is gunned down in the streets by the police, **and nothing is done about it!** The authorities basically sanction it **over and over again**, because it's built right into their system to **need** to have these kinds of things happen, to keep their kind of order, it's that kind of repression that they require for their system to keep functioning relatively smoothly. What a system!

And there are so many things that are wrong in the world. The whole status of women in this country and all over the world—that women are still not treated like full human beings, that they're constantly degraded and dehumanized, treated as play things, as sexual **objects**, as something short of full human beings, constantly raped and battered. And I've said this before: it doesn't matter if it doesn't happen to you as an individual—any time any woman anywhere in the world is raped, battered, pornified, or in other ways dehumanized and degraded, it degrades and dehumanizes **all** women everywhere.

And again I want to say that I really feel that...like BA in the recent Dialogue with Cornel West at Riverside Church talked about...the youth that are being gunned down by the police—these are **our** youth! I feel that, very strongly. And it's just intolerable to have this loss of human life, this loss of human potential, that is just squandered away because of the workings of this system.

It is also intolerable to have a situation where there are endless wars. You can never get beyond this under this system: these wars of imperialism, these armies of occupation, people being put through horrible suffering for the interests of a capitalist class, a tiny sliver of humanity that benefits from this.

And what about on the planetary scale? The environmental crisis is **real**, people! It should be understood as an all-out global emergency. You know, the Earth itself is one thing, it can go on without us, but **human beings' ability to live on this planet** is going to be severely restricted, very soon, if we don't stop completely despoiling this planet and constantly degrading it. And the main reason we can't deal with any of this, fast enough or on a large enough scale, is because of the dominant system that's in place, the dominant form of social organization that's in place. We need an actual revolution to completely dismantle the organization of society as it currently is configured and to replace it with a completely new form of organization that would go a long way towards getting rid of these problems.

Look, also, at the so-called problem of immigration. Why do we even have different countries? Think about it. Why do we have flags and national anthems, and why do we have borders? Why do we have whole populations of people that are pushed around and kept from having a decent life, when all they want to do is work and be productive members of society? Think of all the immigrants to this country who get pushed around, get brutalized, whose families are brutally torn apart, and who get incarcerated, forcibly deported or even gunned down on the border. Do you find that acceptable? I sure don't! What makes Americans better than anybody else, by the way? Personally, I can't stand the American flag, or the national anthem, or the Pledge of Allegiance, or any of these kinds of symbols that proclaim that one country or one population of one part of the world is somehow better than everybody else. That's what's called "jingoism," or "national chauvinism"—that way of thinking is downright nasty and we should call it what it is and refuse to go along with it! We should all be thinking more like citizens of the world and not like Americans. But then you see people stand up in schools and at sporting events—they're standing up for the flag and the anthem, and they're putting their hands over their hearts and maybe even singing along, and often this is being done by people who are themselves being oppressed and degraded on a daily basis by the very system that they are saluting!

It's time to put an end to this kind of stuff. Think about what you're doing, what you're saluting! People need to think more about this, and educate themselves about the true nature of this system. These police murders, for instance: **they're not an accident**. They've been happening for a long time. They happen on a horrific scale. And they keep on happening, because the root of this problem can be found in the very foundations of this system.

The only good thing in this recent period, what you're calling this "moment," is that there's a beautiful new thing that's emerged, which is that people are standing up and resisting in ways we haven't seen in this country in a long time. That's a beautiful thing—the youth and others who stood up in Ferguson, very bravely, and said: NO! we're not gonna take this any more. And the people who came out around the police murder of Eric Garner. And this did involve broader numbers of people, besides the most oppressed people who are most directly under the boot of the police. There were also people from the middle strata, including some white people, who came out and said: We don't want to live in a society where this kind of stuff keeps happening. So that's a good thing, although there needs to be a lot more of that kind of resistance. That kind of resistance is very, very important, and it needs to get bigger and it needs to spread. One of the things a scientific understanding and analysis can tell you is that protests are very good and very important. What's been called "fighting the power" is very important. It builds the strength of the people. It serves notice on the people running society that their crimes are just not going to get over, and are not going to be tolerated any more. And that's a very important part of what needs to happen. But it also has to go further. Why? Because a scientific analysis will also show you clear evidence that the whole way this system is structured, the whole way it's built up at its core, at its very foundation, will keep regenerating these kinds of problems, these kinds of abuses, these kinds of outrages and injustices, over and over again, as long as this capitalist system itself is allowed to remain in place.

Sometimes we talk about **the unresolvable contradictions of capitalism**. If you use science to analyze this stuff, you will increasingly understand that this system cannot fix itself, and that it is not fundamentally capable of correcting these types of abuses. It cannot do away, ultimately, with the police murders of Black and Latino people in this society. It cannot do away with the rule of their enforcers, the brutality of their enforcers, that keeps a whole section of the people down. All this has a lot to do with why BA stresses all the time that you have to understand that this country, this system, **was built on slavery**. It's not just what's happening now, it goes back to the very beginning of this country. The United States got started, got built up, at its very founding, on the **basis** of slavery (and genocide of indigenous peoples), and everything that came out of that brutal beginning has carried over until today, and is a direct root cause of why today you have police enforcers, defenders of this capitalist system, who are routinely gunning down unarmed youth in the streets. There is a direct connection there. Science will show you that this connection is real and objective, and not just someone's subjective opinion or empty speculation. To make such a claim you need concrete evidence—and the evidence is there.

It's the same thing with the question of the oppression of women. It's another one of those profoundly unresolvable contradictions of the existing system. This system cannot ultimately resolve that problem, which science can show has been deeply built into the root foundational structures of this capitalist system as well as those of previous oppressive and exploitative systems going way, way back in time. Yes, there are some women, a few—there are a few sections of women that can be allowed to move up the ladder, so to speak, under capitalism. The same can be said about Black people—a few can be allowed to move on up, to enter the professional middle strata or even become totally *bourgeois*, and you can elect some Black officials to high places and even have a Black president these days. But none of this changes anything fundamentally about the profound and relentless oppression faced by the vast **majority** of Black people in this country, and of other people of color as well. The same goes for women. Literally half of humanity—in other words, women—continues to be kept down, in all sorts of ways, in the U.S. and all around the world, and none of that changes just because you can now have a few female corporate CEOs or government representatives or a few very wealthy *bourgeois* women. None of that changes the ongoing fundamentally degraded and dehumanized status and experience of the vast majority of women here and throughout the world.

Wars of empire—there's another one of those unresolvable contradictions of this system. It doesn't ultimately matter whether, every now and then, even a few individual politicians or other representatives of the ruling class are willing to speak out—even sincerely—against one or another war of imperialist aggression. This ruling class is going to **continue** to wage wars of empire to extend and defend and consolidate their imperialist system. And they will do so over and over again. Why? Because the underlying dynamics of their system **drive** that process, whether any individual politician or other ruling class figure would like it to be that way or not. Do you see? The very machinery of this ghoulish system repeatedly **requires** such wars—for its ongoing maintenance, expansion, and consolidation.

So we have to confront the fact that what we call national oppression, the oppression of minority peoples, and the oppression of women, the wars of empire and the armies of occupation—none of this can ultimately be solved under this system. Science can analyze **why** none of this can fundamentally be solved under the structures of capitalism-imperialism. And this is something that BA has done a lot of work analyzing over decades, really deeply bringing to light why **this system cannot be reformed**, why it cannot just be fixed with a few quick fixes, why you have to have an **actual revolution**, rather than just work for a few little tweaks here or there.

And the same thing is very much the case with the question of the environment, the global environment. Even if you had a bunch of capitalists and other ruling class figures—you know, their government representatives in this country or in other countries—who personally became really convinced that there is an environmental emergency for the planet, and that steps really have to be taken to try to save the planetary environment and prevent all this degradation which is causing critical problems throughout the world—even if some (or even many) individuals in the ruling class became personally convinced of that, and even if they tried to institute a few reforms here or there, they would quickly run up against the limitations and obstacles of their own system! The capitalist-imperialist system is simply not set up and structured in such a way as to allow the kind of radical transformations that are actually needed to resolve the global environmental crisis. Because of the underlying structures and "rules of functioning" of their aggressively competitive and profit-driven system, capitalists are simply not capable, **they do not have the material basis, to actually resolve this planetary environmental problem, with sufficient scope and scale, under the current system**.

This is all very important to understand, and once again it takes science to deeply understand that you can't just "convince" rulers to change, because they are themselves completely caught up in the rules and machinery of their own system, whether they like it or not. The machinery of the capitalist-imperialist system has basic rules of functioning, "rules" which ultimately cannot be changed without changing the type of system we live under. If you don't understand this...if you think the way to change the world...if you think, for instance, that the way to keep the police from killing unarmed Black youth is to just to do a few "reforms," like putting body cameras on the cops, or just do better education and training for the cops, you're going to have a rude awakening, because their system will keep regenerating this form of terror and oppression. It can't **not** do it.

Same thing with all those other situations. If you think that just empowering a few women or girls, in a few instances here or there, is going to get rid of the burden of the systemic oppression of women in this country and around the world, you're deluding yourself. If you think that just expressing the people's will not to go to war is actually going to be enough to ultimately put an end to all these imperial wars, you are also deluding yourself. And if you think that convincing the capitalists that it's better for their bottom line to not degrade the environment so much, or that their children and grandchildren will suffer if we don't actually save this planet...if you think that's going to be enough to solve the global environmental crisis, you're also deluding yourself.

Protest? Yes. Definitely. Protests are very important. It's very important for masses of people, here and all over the world, to make clear that they won't tolerate and be complicit in and accept any more these egregious abuses and injustices. It is important to say: **NO**, we won't take this any more. As I've said before, it is part of building the strength of the people. But you have to go further and understand that there are built-in contradictions within the way economies and politics are set up under certain systems and that those underlying contradictions—there are clusters of them that lead to horrible injustices and abuses—are simply not resolvable by the capitalists, under a capitalist system. You need a different economy, you need a different ideology, you need a different worldview, you need different social objectives. You need different forces coming to the fore to implement that. You need state power. The people have to organize themselves for an actual revolution. And, you know, in the course of just this interview, I can't really go into all the patterns that prove that those underlying contradictions of this system cannot be resolved by the system, but there is accumulated evidence, including very much spoken to in the extensive body of work of BA, that has been developed over decades, over more than 40 years. The work has been done, the work is deep and profound, it is scientific, it is methodical and systematic. And people should critically examine it, they should engage it, they should study it. It should not be dismissed by anyone superficially. It is getting at the underlying deeper problems and corresponding solutions.

I'll just say this, and then I'll stop for a minute on this point [laughs], but one of the most encouraging things about science, too, is that it shows you the potential for positive change, how we could change things in some really good ways. If you don't have science, you're kind of bopping around in life, running into problems, maybe solving a small problem here or there, but more problems keep coming up, and you don't know what you're doing, basically. But with science you can systematically figure out not only the source of the problems, but also what the basis is for positive change. One of the things that people don't understand very often is that **the basis for the revolutionary transformation of a society, of a social system, where that basis is located—it actually resides right within the contradictions of the system**. In fact, right within those contradictions I was just talking about—the really big ones that this system cannot fix itself, that it cannot ultimately resolve. The fact that they can't resolve these big things and that they keep driving people into the ground in different ways, actually creates the conditions which move in the direction...actually creates **the basis** for people to be able to work on those contradictions, to bring the people forward, in the thousands, in the millions, to move towards the ability to organize for an actual revolution, and build a new society on a whole different basis. That won't resolve every problem overnight, obviously. But many, many of the big problems can be resolved to a great degree, thanks to science, and thanks to the conscious initiative of people organizing themselves collectively for an actual revolution.

Personal Experience and Development: Intellectual Training and the Joy of Scientific Wonder

Question: OK, well I wanted to move to talking about your own experience and background and development, because I think, frankly, it's fascinating, and I think it would really be interesting and full of lessons for a lot of people, and I thought it would be fun to get into that. So, to start out with that, your background is as a trained natural scientist, and I think one thing a lot of people might be interested in is, how did a trained natural scientist become a revolutionary communist? Maybe you could speak to your journey and process there.

AS: Well, I'm not sure of all that I should get into, but in terms of my background, I was first of all trained as an intellectual. I was privileged to have a very broad, liberal arts education, and I was specifically trained as a professional biologist in the field of ecology and evolutionary biology. And a great joy of my life has been to be able to take up work in the natural sciences, in the natural world. From early childhood, I've always been infused with a sense of wonder and curiosity about pretty much everything [laughs], and particularly in the natural world. So, I really enjoyed functioning as a scientist and taking up scientific methods to go out and explore, to go out and investigate, to try to learn more about reality, to learn more about the dynamics of different natural ecosystems, whether you're talking about rain forests or desert ecosystems, or coral reefs, or temperate forests. I could go on and on [laughs] and talk about all the fascinating dynamics and interactions between plant and animal species that I had a chance to explore in these settings. And my life could have gone on that way, very easily, from that point on.

But I'm also a child of the '60s. Socially, I've been formed by things like the movements of opposition to the imperialist war waged by the U.S. in Vietnam. And, in that period, I also developed consciousness of the lopsidedness in the world, wherein the relatively high standard of living in a country like the U.S. was in sharp contrast to the standard of living in most Third World countries, which I got to see firsthand in the course of my work. And that contrast, I came to understand, was one where that high standard of living in the U.S. was built on the backs of the people in the Third World, and that was part of my developing social consciousness. And then, inside the U.S., I became acutely aware of the systematic and systemically foundational nature of the oppression of Black people in particular, which horrified me; and, during that whole period, I was inspired by the struggles of the civil rights and the Black liberation movements. When I was a student, I joined in, in support of that. I joined in demonstrations against the Vietnam war. And, of course, the woman question—that was also, in that period of time, a burning issue in terms of people really starting to dig into the status of women in the U.S. and around the world, and why were there such conditions of systemic oppression. There was the question of reproductive rights, the right to abortion. I became very clear on the question that if a woman did not have the right to control her own reproduction, to determine when to have children, or whether or not to have any...when a woman does not have the right to make those kinds of decisions, that is essentially a form of enslavement. I became very conscious of that, way back when. And that was part of the formative experiences for me, too.

And even back then, especially because I was trained as a biologist, the questions of environmental degradation and the loss of species diversity, the deforestation of the rain forests around the globe, and problems such as these—I was acutely aware of all this, and acutely aware of how fast that destruction was taking place. And some other scientists and I were very concerned about what that was going to lead to, in terms of possibly making the planet ultimately uninhabitable for humans, and, in any case, certainly leading to a great loss of natural beauty and resources all around the world, all because of the depredations of capitalism. That was one thing that, in the '60s, many people, even at a sort of primitive level, came to understand. People sometimes called it "the military-industrial complex," or "the establishment," or whatever, but at least many people knew it was **a system**, and many people increasingly understood and discussed basic concepts like capitalism and imperialism, and more and more people were coming to understand, at least in a simple way, that this was the nature of the problem that was underlying all these social problems.

So these were some of the political and social formative experiences that I had, at the same time that I was very much involved in the scientific sphere. And, I think that, because of my involvement in and my training in the scientific sphere, I've never been particularly drawn to superficial analyses, or analyses of things that were just based on what people were thinking. **I've never been too concerned about what people think, in the sense that I don't assume that just because a lot of people think something, it's necessarily right; nor do I think that, just because only one or two people think something, it's necessarily wrong.** I rely instead on scientific methods and accumulated evidence. I don't evaluate things based on popular consensus ("what most people think"), which is all too often wrong and out of step with actual reality. My scientific methods make me very much a critical thinker—I've been trained in critical thinking since early childhood, and it's a very important part of who I am, to be a critical thinker. At the same time, with scientific methods, I want evidence, I want material evidence. If I'm trying to understand something in the natural world, I don't want somebody's opinion. That might be a place to start: Somebody might have an idea, an intriguing question, an opinion. They might engage in creative speculation. But then, you've got to take it somewhere, turn it into some kind of project, experimentation, go out into the real world and investigate it. It's not enough to state your "opinion," or even the "opinion" of lots and lots of people. An opinion that is not backed up by evidence might reveal something about **you**, but other than that it's pretty worthless! [laughs]

So go on and **investigate the real world, thoroughly and systematically, and provide evidence for your analyses and conclusions**. Look for patterns, repeated patterns. Look for evidence that comes from different directions. Keep an open mind and work with integrity to determine, on the basis of evidence, whether something (in nature or society) turned out to be as you initially expected it might be, or perhaps turned out to be something altogether different.

Coming out of the '60s, there were many different political trends and movements and organizations, and most of them, frankly, didn't inspire me at all. They seemed to be kind of narrow and mechanical, and often very narrowly economist, where they were just trying to get somewhat better living conditions, or better working conditions, for some people, but they didn't really get to the root of the big problems. Or some other kinds of movements were kind of steeped in nationalism or feminism or something—you know, the beginnings of identity politics. Even then I really wasn't interested in political philosophies and movements that could not encompass **the whole range** of the key problems of capitalism and imperialism. I had come to the conclusion that the enemy was imperialism, and I wanted to know who could deal with **that**.

And when I was in college, I read Mao, and got inspired by the revolution in China, and tried to learn some more about it. That was very formative for me, too. And later, as I basically got to know people in the precursor of the RCP [Revolutionary Communist Party], the RU [Revolutionary Union], and then what became the RCP in the mid-'70s, even in those early days, I encountered the leadership of Bob Avakian through some of his works, in particular through some of the insightful analyses that he was making of, for instance, the Black national question in the United States, and the arguments he was making about the direction things needed to be going in. And there were mistakes in some different areas, things were still pretty primitive in some ways back then. But he already stood out to me as somebody making substantial analyses, applying more rigorous scientific methods and making deeper analyses of societal problems than people who seemed to be working just on surface phenomena. He was going deeper and digging for evidence and his more scientific approach to the problems of society already stood out to me as being different from the ways most people in the "movements" of the day were approaching things, and this really appealed to me given my science background. As I said before, I've never been impressed by populism, by the notion that what's popular, or what most people think, is what things should be based on. That doesn't carry any weight with me. I want to identify methodically what is the cause of the deep problems in society, and I want to try to work with people on solving those problems.

So I guess I approached all this very much the way I would approach projects in the natural sciences. Some of my favorite experiences in the natural sciences involved working collectively with other scientists to wrangle with questions—asking a lot of questions about something that's not well understood yet, and then batting around how we might tackle the problem, how we could learn more about it, and dig deeper. Are we getting convincing evidence of the characteristic features of a phenomenon? Are there experiments that we could do, ways we could work on reality, that might bring to light some of the underlying patterns and that could either reinforce or challenge our current understanding, and further develop it? That would reveal the underlying material basis for how some things have changed in the past, are changing now, or could change in the future?

Coming out of the '60s, I was exposed to the concept of dialectical materialism and realized that the analysis of **underlying material contradictions** could be applied to any aspect of the material world, and is in fact a key method for deepening our understanding of both the defining features and characteristics of a thing or phenomenon and its patterns of motion and development. And this applies in both nature and society. To this day, I walk around in all sorts of different natural environments and what do I see? I see contradictions. I see contradictions everywhere! [laughs] That's how I see the natural world. If I'm looking at a hummingbird pollinating a flower, I'm seeing it as a contradiction, I'm thinking about the contradictions.

When I say "contradiction" here I am not talking about a conflict or an antagonism. Everything is made up of contradictions but not all contradictions are antagonistic. In the sense I am using it here, contradiction is just a "rapport," a dynamic relation or interaction, for instance between a hummingbird pollinator and the flowering plant it is pollinating. And that particular dynamic contradiction is itself situated and playing itself out in the broader context, and in dynamic interaction with, a much larger ecosystem (perhaps a rain forest, or maybe just a backyard garden) which is itself made up of a great many other dynamic particular contradictions within and between the many different elements that make up that broader ecosystem. And then of course there are always lots of dynamic contradictions that come in and impinge on things from outside a particular system, often pushing change in some entirely new directions. So, if you're really trying to understand a process, any process, there are questions to consider about different levels of organization of matter, about differences of scale; there are questions to consider about both the internal contradictions within a process or within a thing that define its principal characteristic features as well as some of its pathways for change; and then there are also those external contradictions that can come in and impinge on the whole process and push things in new directions, though always on the basis of those underlying systemic contradictions. I'm not trying to get into all this too much right now, but this kind of dialectical materialist thinking and approach is critical for doing good science, in both the natural and social spheres. And so, yeah, this is how I try to think about things whenever I'm out in a natural ecosystem. I'm asking questions in my mind, and exploring and thinking about these things: what are the underlying causes, how do the underlying contradictions inside a system or an entity actually provide the material basis for that thing, that particular entity, or that particular system, to change over time? I understand that the fundamental basis for a thing to change is contained within that thing, in interpenetration with its external environment. And I'm interested in things that happen on the edges, and on the borders of things. If I'm walking through a forest and I come to a clearing in the forest, I immediately start thinking about this—the particularities of the interactions of edge species, the particular assemblages of plant or animal species in those particular habitats, the dynamics that are taking place at the borderlines between the forests and the clearings.

But I don't think we can get into all that too much now. [laughs] I'm trying to make the point that there is a continuum in my thinking between how you approach matter, matter in motion in the material world of what we call the natural world, any aspect of the natural world, and how I look at the material world of social reality, the way societies, human societies, are configured. I had some training, a bit of training, too, in cultural anthropology, and I've always been interested in the history of human social systems, from foraging societies to other kinds of societies—agricultural societies, advanced industrial societies—societies organized on different material bases, on different economic foundations, and what effect these underlying forms of economic organization had on the ways of life of people, and their traditions, and their cultures, and what were considered norms and what was considered right and wrong, and so on, and how that could change over time and depending on the social system underpinning it.

So, there's continuity between these different areas of interest in my life as an intellectual. And I think it's not an accident that what I've found myself most drawn to, even at an early point, was the more scientifically oriented approach to the transformation, the radical transformation, of society. I became convinced early on, back in college days, that the problem was imperialism and the solution was revolution, and moving towards some kind of a socialist society in the direction of some kind of communist world. Well, I didn't necessarily have a very deep understanding of how to go about it, or what that might mean, but I could sort of see in a basic sense that that's where things needed to go. And I've never been diverted from that basic understanding in the time since. But what's happened is that I have been able to learn—including, thanks to the work developed by BA, I feel I've developed a much deeper understanding of those underlying social dynamics, and wherein lies the material basis for positive radical change.

Coming to See Capitalism-Imperialism as the Problem, Being Drawn to Communism

Question: What was it that...obviously it wasn't one thing, and people spontaneously don't see that the problem is this capitalist-imperialist system, or that the system can't be reformed, but you talked about how, when you were in college, you came to see that the problem was the system, even if it was on a basic level that you understood that. How did you come to see that the system is the problem, and that the system needed to be gotten rid of?

AS: Well, I can't take credit for that myself as an individual. Like most people, I drew inspiration and insight from reading, from what other people had worked out. And this was a time of great ferment all around the world, including intellectual ferment and intellectual engagement on big questions. So, for instance, this was the time of the Vietnam War, and people were asking questions: Is this a just war? Not all wars are unjust, but is this a just war? If not, why not? What is the cause of this war? Why is the U.S. going all the way over there to bomb a bunch of Vietnamese people into oblivion? Why is that happening? Whose interests does it serve? And there were people around the world writing about imperialism, writing about colonialism in Africa, in Asia, in Latin America. One of the things that was important in that period is that, all around the world, there was more consciousness about the role of U.S. imperialism, the negative consequences of that, and so you had people around the world denouncing—you had mass demonstrations against—U.S. imperialism. You know, an American president would travel to Latin America, and people would turn out on the streets in cities all across the region with signs that said "Yankee Go Home!" and making it clear they didn't want the intrusions of U.S. imperialism in their countries. That kind of resistance to U.S. imperialism was very prevalent at that time. And, of course, it also made a big difference that there was a genuine socialist state in the world at that time. I'm talking about China, which, at that time, was a genuine socialist society, led by Mao and the Chinese Communist Party. That made a big difference. Not just for the people in China but for people all around the world who were challenged and inspired by this. It served as a model. It was a very different kind of country. It was a socialist country, and it was a Third World country that contained about a fifth of the world's population at the time, and it had broken radically with the previous system of organization of society and was in the process of undertaking this major social experiment to build a completely new kind of society, on a new economic, political and ideological foundation. That was very exciting to learn about, and many people, students and others, did try to study it and learn from it.

But then, later on, when the revolution got reversed in China in the late 1970s, after the death of Mao, this posed some new questions and it was very confusing for a lot of people. First of all, it became very important to understand that there actually HAD been a reversal of the revolution in China, that the whole society was being forcibly diverted from the socialist road and put squarely back on the capitalist road. First of all, you had to understand that this was actually happening (many people refused to believe it), and then you also had to dig into **why** this was happening. Mao himself had actually warned us about this possibility—before he died, he had worked out some very important concepts about the need to "continue the revolution under socialism," and he repeatedly stressed the reality that under socialism the bourgeoisie could be found "right inside the communist party," and he urged everyone to take part in mass campaigns to compare and contrast opposing and contending lines and programs and help wage sharp struggle to keep society moving in the right direction.

Ultimately the revolution in China was in fact reversed, and the whole of Chinese society unfortunately was wrenched back onto the capitalist road, evidence of which can be seen in the many horrors once again plaguing that society in recent decades. This was a tremendous loss, not just for the people of China, but for people all around the world, and we are still feeling the effects of that loss right up to this day. But many people (and I count myself among them) did learn a great deal of very important lessons from this whole experience—lessons that will never be forgotten. And these lessons will be applied to taking things even further, and building things on an even better foundation, the next time around. And I have to say that, once again, Bob Avakian really stood out in that whole period in his ability to scientifically analyze (correctly) what was going on in China—at a time when most revolutionaries and communists around the world were lost in a daze of confusion and unscientific denial. This was very valuable guidance and leadership, for anyone who cared enough to listen. And BA didn't stop there. He didn't just analyze what had happened in China. He dug further into why the revolution had been reversed and capitalism had succeeded in regaining the upper hand in China (and in the Soviet Union as well, back in the 1950s), and he went to work on deeply studying and sorting out the great accomplishments of those revolutions from their secondary shortcomings and deficiencies, with a particular focus on problems of method and approach. Avakian's *new synthesis of communism* is the direct result of decades of systematic work on these very questions and, in my opinion, it represents a tremendous forward advance in the development of the scientific theory that is needed not only to correctly guide the next rounds of revolutions but also the construction of new societies that most people would actually want to live in.

But getting back to your question, my point is that in the 1960s and '70s there were all sorts of people, in the United States and all around the world, who were studying Marx, and Lenin, and Mao, and trying to learn from these theoreticians of the revolution in different periods of history. People were studying political economy. And people were talking to each other about such things. College students in particular, and other intellectuals, could often be found passionately talking to each other about big social issues and exploring questions such as: What is commodity production? And how does capitalism work? And is it possible to actually mitigate some of the problems of capitalism? Or does this system as a whole really have to go? Really big questions, with really big implications. I can't take it too far right now, but revolution really was in the air back then, for real. Many people had different kinds of analyses, and were obviously not all on the same page about what needed to be done, but a great many people were at least recognizing that a lot of the most outrageous and egregious abuses of this system—on the national question, on the woman question, on the environmental question, on the question of imperialist war—had roots that could be traced right down to the mode of functioning of imperialism and to the interests of the capitalist class running this society. That understanding was often a bit simplistic and primitive back then, but it was basically on track.

So people talked a lot about such things. They read, they studied. You know, people sat around talking about it late into the night. When people went looking for organization and leadership, they checked out different groups. Like I said earlier, I myself checked out a bunch of different groups, but I wasn't very inspired by the ones I initially encountered. There were people who clung to pacifism; there were others who started to recognize that you would need to get to the point where you could have an actual revolution—to actually overthrow the existing system. Some people were impatient and fell into various **adventurist** shortcut ways of thinking and acting, as if you could somehow spark a revolution, and carry it out, on the basis of just a few handfuls of dedicated people. Some of these people were brave, but they were also very unscientific, and I always thought such approaches were incredibly irresponsible, that you would just get people crushed, without a real chance of actually transforming the society. But, on the other hand, I also was not at all attracted to **reformist** schemes and ways of thinking. I didn't like those who proposed only superficial band-aid solutions to the really major problems of society. I was looking for deeper, more fundamental answers and solutions. And I think that this very interesting material that was being developed by Bob Avakian, even in the 1970s, already kind of stood out. It was different. Avakian seemed like a different kind of theoretician than what I had previously encountered. He came across as someone who was serious and had some theoretical substance, but who at the same time was also working very practically on the concrete problems of the revolutionary movements of the time. So the connection of theory and practice there was something that also attracted me.

And the other thing I would say that attracted me was that this (the Revolutionary Union, which was soon to lead to the founding of the Revolutionary Communist Party, in 1975) seemed to be the organization and the leadership that could really bring together a wide variety of people: people who had absolutely no education, who came from the hardest streets of the inner cities of this country, working together with people who were college professors or college students, or little old ladies, or retired people, or whatever—the variety of people that could be brought together appealed to me. The many different nationalities, the different ages and backgrounds, and so on. That was very exciting to me. Early on I found myself in meetings and discussion groups, and so on, with that wide variety of people. And it was wonderful. It was wonderful. It was kind of giving you a taste of what the future society could be like, because all sorts of social divisions were being overcome by these diverse people who were coming together to really try to grapple with the problems of this society, and of how to make an actual revolution to get past it and get to a different kind of society.

Getting Clearer on the Need for Revolution—Breaking with Wrong Ideas and Illusions

Question: In the process of coming to see the need for revolution and communism, what were some of the key previous ways of thinking that you had to break with?

AS: Well, let me see. One thing—it didn't last long, but when I was in high school, I briefly went through a little bit of a pacifist stage. I remember making little peace symbols out of copper wire. [laughs] Look, like most decent people, I'm not inclined to simply accept human suffering and death and destruction—if this could be avoided, I'd say so much the better. But I came to understand, even early on, the nature of the system that dominates this society and that causes so much exploitation and oppression, and I came to understand the tremendous violence of that system—violence that is perpetrated on the people on a daily basis. All you have to do is look at all the police brutality and murder that goes on, that everybody's been talking about lately, and which has been going on for a long, long, time. That's one example of it. But there is also what they do in the course of their imperialist wars—and they're ruthless. I mean, these are very, very violent people and very, very violent institutions. And, yes, they're very powerful. They have tremendously sophisticated and developed weaponry and military forces, and so on. So you'd have to be very non-materialist and very naive to think that you could just, at some point, politely say to them: "Excuse me, but could you please just step aside and let us run society in a more reasonable and rational manner that would benefit most of humanity? Oh, and by the way, you and your way of doing things? You're out!" [laughs] To think that they wouldn't come back at you with tremendous violence, with everything they could throw at you—you'd have to be very naive to think that. So, this became very clear to me. I don't know any revolutionary communists who are thirsting for blood and carnage, or any such thing. You are talking about decent people, who are not cold to the reality, who don't fail to understand what it means for people to suffer and die and to lose close friends and family. But yes, I broke with pacifism, even in high school. Like I said, it was a very brief phase, because in those days people were talking about what it was that imperialism was actually doing here and around the world, so you got to see and hear about their tremendous brutality, the tremendous violence they were routinely perpetrating. People were talking about it, and were willing to investigate it, and were willing to share that knowledge with each other. They were not just trying to cultivate their own gardens.

So again, this was a time when there was a lot of mass discussion and debate around such things as the question of reform versus revolution. Would it be better to try to "work within the system," or outside the system? This was a mass question in society at that time. Could you change things through elections? Could you just try to find more progressive candidates? Is that the way you should try to change things? Or did you need to recognize that the system itself was functioning on a basis that could not really accommodate to a new and more just and equitable way of life, and that could not deal with getting rid of all these injustices and outrages and violence, and that it would therefore have to be forcibly removed as a system in order to clear the way for a new kind of society?

Now, people in those days in the U.S. certainly didn't know too much about how to go about making a revolution. At the time I personally certainly didn't know anything about revolutionary organization or revolutionary strategy. When I was first being affected by these social movements and developments, I had never even heard of the concept of what's known as a Leninist party, a vanguard party. I had no sense of why one would even need such a party in order to be able to carry out a revolution. And there were a lot of other things that even revolutionary-inclined people just didn't know anything about, but that would be essential for making an actual revolution in a country like the U.S. For instance, how would you go about uniting people very broadly, but still manage to maintain strategic focus on preparing minds and organizing forces for an actual revolution aimed at dismantling the existing system and at setting up the basis for a socialist society? What forces should you involve? What forces should you rely on? There were many, many such questions. What kind of stages might this go through? And how would you even begin to set up a new society? Again, things were very primitive in some ways back then, and there were more questions than answers, but many people were actively searching for those answers. And what was impressive was how many people cared, how much they cared, and how many people were willing to sacrifice a lot of their own lives and a lot of their—frankly, a lot of their own happiness, or stability, or safety, or things like that. And that was inspiring, too. People in large numbers were willing not only to dream of a better world, but also to take steps and act in accordance with those dreams.

Personally, I would say that the other thing I had to break with was...look, I didn't come from the hard streets. My family was always strapped for money when I was growing up but, for a number of reasons, I was able to get a very fancy education, and therefore I had a lot of entrée, or access, into a very privileged world of intellectuals. And that meant that there was a basis for me to end up having a pretty comfortable professional life, doing all the things that I really enjoyed doing, and making a living at it. I was lucky enough at a very young age to be able to have a lot of very positive experiences that way: to travel internationally, to conduct scientific experimentation, and basically to thoroughly enjoy myself. But at a certain point, I really had to confront the question of "self," and how much was I going to remain on a track that encouraged and basically promoted and prioritized my own well-being, versus how much was I going to dedicate myself to trying to make a better world for humanity in general, and work on relegating "self" to a more secondary position, and no longer proceed from prioritizing just my own personal needs.

And look, I suspect there are quite a few people today who are like I was back then. People who have a lot of potential, who could make a lot of contributions to the revolution, but who still have some trouble with the notion of subordinating "self" to something larger than themselves and putting first things first, on the right basis. Especially given the prevailing "me, me, me" culture of today! But I guess, first of all, you have to care, right? That's one thing that in my own experience I could never quite shake off: I actually did care a lot about the outrages, the injustices, the tremendous unnecessary suffering people were subjected to in both the United States and in the Third World. Now, caring, in itself, that's a good start, but it's not quite enough. The next question becomes: Are you going to do some work to get a deeper scientific understanding of why all these outrages keep happening? Why can't we get past all this? Why do these same problems keep coming up over and over again? Why can't we get to a better society, to a more reasonable and rational society, that would actually benefit the vast majority of people?

Then, once you start finding the scientific answers to those kinds of questions, a new set of moral questions comes up. OK, now you know enough: In at least a basic sense you know what the source of the problem is; you know that the system can't be reformed; you know that it's going to take a revolution, and that revolution is a complicated process; and you also know there aren't enough people who have this understanding yet, that there aren't enough people who are part of this process, and that a whole lot more people are going to need to get involved for there to be any chance at succeeding in making an actual revolution. So then that poses a moral dilemma, a dilemma of conscience. You get to that point, and you basically have two choices: you can either look at yourself in the mirror and say: I know too much to turn away, and I really have to become part of this; or you say: Well, you know, I kinda like my life, thank you very much, and I think I'll just go on and do what I feel like doing as an individual, and turn my back on the people who are suffering under this system.

Once again we are at a time when a lot of people should be asking themselves these types of questions.

Question: I think this relates very much to what BA talks about in terms of the head and the heart. There is the scientific understanding that the world doesn't have to be this way, and it could be radically different. And then it seems like—another thing your own experience points to, including for intellectuals in particular, is having to make the decision—at a certain point you made the decision to fully give your heart to humanity and to the revolution and to the masses of people. You could have had, I'm sure, a career as a natural scientist...

AS: Well, I did.

Question: Well, you did. But I mean you could have continued to focus on that career. Right? You could have kept going that way. And you made a decision at a certain point to give your heart to the revolution. So how did you make that decision to give your heart to the revolution?

AS: Well, I think it's what I was just saying: realizing that I knew too much at that point to turn away from what I understood. What I understood on a scientific basis. Also, having gained some sense of the revolutionary possibilities. I never thought that the revolutionary process would be an easy process. I always expected that there would be sacrifices and risks. Like many people who came out of the '60s, you expected that you might be jailed, you might be killed—just for opposing U.S. imperialism. I mean, look at what happened to the people at Kent State and Jackson State, for instance. They were college students—but that didn't save them. When you get to a certain point, if you have a heart and you care, and in addition to that you also have some scientific understanding of problems and solutions, then it becomes pretty difficult to live with yourself if you turn away from all that. Because then every single time you open a newspaper, or you turn on the TV or something, and there's Trayvon Martin dead in the street, or Eric Garner, who can't breathe, being choked by the police right on video, or Mike Brown gunned down so brutally—and all the Oscar Grants and all the Amadou Diallos and all the Sean Bells. And they stay with you, you don't turn away, and you don't forget them. There are so many outrages like that, and when I see those things, **I feel like any one of them is enough for me to want to make a revolution**. Any one of them is enough of a reason! Because I do understand: It's not an accident, it's not an anomaly, it's not something that "just happened" because of some individual rogue cop or something. I understand how all this is systemic, it's built into the very fabric of the capitalist-imperialist system. That's why these things keep on happening. And it's the same thing every time a woman or a young girl is cast out by her family for being pregnant, or a woman becomes pregnant and seeks an abortion but she can't get one because there are no longer abortion clinics in her state, or she has to travel many miles away, and she ends up not being able to get an abortion and is forced, literally forced, to give birth to a child she never wanted or is simply not ready to raise and take care of because of her circumstances. I look at all that cruelty and I recognize it as a form of slavery. A woman who is denied the right to control her own reproduction is reduced to the status of a slave, and all women everywhere are objectively degraded by this. So again, all these outrages, any **one** of these outrages...that's enough for me to want a revolution and to get serious about it.

Or every time this system uses the death penalty against people—you hear about people who've been convicted who were clearly innocent, and who've been thrown into the dungeons of the prison system for decades, or are executed; or people with obvious mental health problems who are executed. Any one of those kinds of things is enough.

Every time they turn away people at the border, or they deport people and break up families, or they gun down people on the border, and label people as "illegals".... Any one of those examples is enough for me to want a revolution, and to want to work for it.

Every time I see a homeless person, trundling along, trying to find a place to sleep for the night, because they can't...because in this incredibly wealthy society, there's not even a place for them to get basic shelter! Or people are going hungry. BA sharply called this out: **why isn't there a right to eat?**

Or when I hear about things like the U.S. sending its drones and its bomber planes to countries in the Middle East, dropping bombs on people, wiping out civilians, I don't think about, Oh, what fancy technology they have, or how clever those drones are. I'm thinking about bodies exploding and brains splattered, and families broken up, and suffering horribly. Any one of those things is enough to want a revolution.

When I think of women throughout the world and the sex trade, and the promotion of pornography, where generations of boys and young men are being trained to think of sex basically in rape-culture terms, and they have no idea—nobody seems to have any idea any more—of what really good sex is, or feels like, or how to have decent relationships. And there's this constant promotion of the degradation and dehumanization of women as mere sex objects and a widespread and worldwide trade in young girls and women as literal sex slaves. When is enough enough?

So any one of those stories, that you can find in any daily newspaper, or on the television news or on the internet, is enough of a reason for me. And multiply that by millions of times. But it wouldn't be enough if I just thought, this is horrible, this is tragic, this is terrible. If that's all I thought, or understood, then frankly I would probably get pretty discouraged and depressed about it all, and I would probably kind of turn away from it. Maybe I would just stop reading newspapers, or watching the news or checking things out on the internet, you know, because it would be so discouraging. But, I don't turn away from it, and I don't get numb to it. And the **reason** I don't turn away is because **I do understand what the scientific evidence tells us about what these problems are rooted in, in terms of the fundamental form of organization of a capitalist-imperialist society**. I understand that these things are direct outgrowths of that particular form of societal organization. I understand that, in past times, human beings found very different ways of organizing their societies (not that they were any great shakes, or free of oppression because they weren't), and this reminds me that human beings could once again re-organize their societies on a completely different basis: one such radical reorganization would be to replace the existing capitalist-imperialist form of society with a socialist society, that is in turn moving in the direction of an even more fully emancipatory communist society. And that would be, I am absolutely convinced, a far better world, not just for a handful of people here or there, but for the vast majority of human beings across the planet.

So that's what keeps me going: the understanding of the problem, and the understanding that there is actually a material basis, in the existing relations of society, to transform things in that direction, toward revolution and socialism, and ultimately a communist world. It's not gonna happen all by itself. It's not like the system's going to collapse by itself and then one fine day we'll wake up and say, oooh, I guess capitalism collapsed, so now we can build up a new and better society. No. It is going to require conscious human intervention. It is going to require people banding together to consciously work on the problem, to work on those contradictions, to develop a process that creates new conditions that will ripen towards being able to have a revolution. It is going to require that. But it is possible.

And I'd much rather live in that kind of new society, and any sane person should want to do that, too.

Question: I think part of the point is that a lot more intellectuals and scientists need to do what you did, and give their heart to the revolution and to the masses of people.

AS: Well, obviously I agree with that, because you need more and more people to join in the revolutionary process. But I don't want to make it sound as though it's something that people have to do right when they are first learning about it. If people are newly checking this out, it would be pretty daunting to think like, Oh my god, I can't even look into this, I can't even learn about it, because these people are gonna put pressure on me and ask me to become part of everything right away, and to commit my life to this, or something. And I think it's important to understand that **people can be part of the process at different levels and to different degrees**. Just make a start. There's a place for everyone. There's a place for people to come into things, and learn about it, to just check things out, and become informed. That's the first step.

Educate yourself, become informed, and go regularly to that revcom.us website. Definitely check out BA's works. Talk to people. Become part of the struggle around one or another issue of importance to you. Keep your ears open, study and learn, and evaluate things as you see what's before you. You shouldn't feel like you have to make a life commitment to things to just start getting into it. And the same thing in terms of being part of the process. There are many, many different ways that people can contribute to the process. Some people will devote their entire lives to it, and it will become the primary focus and priority of their lives. Other people will contribute to the extent they can in different ways: Some people will contribute money, some people will contribute support in different ways, some people will participate in helping to spread the word about one or another initiative, or they'll contribute in other ways.

Once again, there is a real need for more and more people to join the revolutionary process, and a great need for growing numbers of people to actually dedicate their lives to this, in the fullest sense, but I'm trying to make the point that I don't want people to feel like you have to go from zero to 60 right away. I'm a scientist in my approach. I don't think anybody should jump to something just on the basis of, say, a few people talking about it. Come in and do the work, and check things out, and learn about things, and be part of the process. Ask your questions, learn more, ask more questions. Compare and contrast what you encounter with other viewpoints and approaches and methods. Above all, check it out in relation to reality, and see if it seems to correspond to reality as it actually is. And then act accordingly.

On Attending the Dialogue Between Bob Avakian and Cornel West

Question: Well, you were starting to highlight the work of BA, Bob Avakian, the Chairman of the Revolutionary Communist Party, and that's actually where I wanted to turn next. A theme that I thought we would focus on a lot in this interview—because you're someone who follows the leadership of BA, you take his work as the foundation and the framework of your own work, and you're an ardent fighter for BA and for his work and leadership—so in the course of this interview, a big theme I thought we would focus on is what possibility this work and leadership opens up for humanity. But maybe, as one way to get into that, I wanted to ask you specifically about the recent Dialogue between BA and Cornel

West, which you mentioned before. You actually had the chance to attend this Dialogue. The Dialogue was entitled "Revolution and Religion: The Fight for Emancipation and the Role of Religion," and it was held at the Riverside Church in New York City, this past November 15 (2014). So there are a few questions I wanted to ask you about that. But as a starting point, maybe you could just give people a sense of what it was like to be there and to experience that Dialogue.

AS: Oh god (!), it was really great to be at this Dialogue. I'm so glad I was able to be there in person, and I'm also so glad that the live stream is available for anybody who wasn't able to be there. I would encourage people to go to the revcom.us website, and you can access it right there and experience the whole thing. And I am really excited that a high quality film is being made of the Dialogue, which will soon be available as well.

I don't even know where to start. It was like there was magic in the air. It was one of the most hopeful things that I've seen in a very long time. I think it was historic in many different dimensions: in terms of the topic that was approached; the people who were involved in it, the two speakers; the moment in time. I felt like I was able to see a great demonstration of morality and conscience applied to dealing with the problems of humanity—that both speakers stood out this way.

I am sick to death of the culture that prevails so much in this society today that is all about self-involvement and self, individualism, and so on. In contrast to this prevailing culture of basically small-mindedness, self-centeredness, selfishness, whatever you wanna call it, here were two people, Bob Avakian and Cornel West, who have different views on many important questions, but they came together to speak to the people together in a way that was projecting tremendous morality and conscience, a tremendous amount of social responsibility. And I thought, yes, please, promote this, let's have more of this. I thought it was a wonderful example of how you could have principled differences—you could have differences and debate and discuss some of those differences in a principled manner, but draw out the points of unity. They were both so generous in spirit, and part of **why** is because they're not focused on self, neither one of them; they're different people, but one of the things they have in common is that they are both trying to think about the conditions of the oppressed and all the horrors that are visited upon so many people on a daily basis in this country and throughout the world...and what could be done about that.

And, in Bob Avakian's case, he's been spending his whole life, decades and decades, developing work that is deepening our understanding of why these problems are not just accidental, or periodic anomalies—how they actually stem from, originate in, the deeper structures of the system, and why it's the system itself, the system of capitalism-imperialism, that has to go, and be replaced with a completely different system, before we could really emancipate humanity. He brings that to life, and he's dedicated his whole life to studying and bringing out to people, in a very scientific way, in a very rich and developed way, why that is the case, what is actually needed, what kind of revolution is needed, what is the strategy to actually be able to get to revolution, how can we actually have a serious strategy for seizing power, for dismantling the existing state apparatus of capitalism-imperialism, and replacing it with a new state apparatus of socialism, socialist institutions that move in the direction of a communist world that would be a genuinely emancipatory journey for the majority of people. He's done a lot of very serious scientific work on this over decades. Has he ever made mistakes? Of course. Will he make more mistakes? I'm quite sure—everyone does, you know. The point is that he's willing to examine his own mistakes and the mistakes of others throughout history, throughout the communist movement, and in what's been done by other forces in society—constantly being like a good scientist who is actually willing to do critical examination of all of this to try to figure out what's right and can move things forward in a good way for the majority of people, and what's wrong and can actually take things in very bad directions. And even when the mistakes come from the historical forces of the revolutionaries or communists in this country or around the world, he's willing to examine that. And so, because of that, you feel like you're in the presence of a real scientist who's actually going to work and has been working for decades. It's like a very advanced scientist who is at the top of his field in terms of analyzing empire, in terms of analyzing the sources of problems and the alternatives and how to get there, and what pitfalls to avoid, what are the dangers, what are the wrong kinds of thinking that people can fall into and do fall into. You don't have to agree with everything, but you can really feel like you're grappling with a scientist who's being serious about this, and whose heart is with the people.

And what you see with Cornel West—and BA pointed that out in his part of the Dialogue—you see someone who is a very wide-ranging intellectual who's studied many different questions, and who is very concerned about the history of oppression, but who also recognizes that it is not enough to just be an intellectual behind closed doors who thinks about these things...it is important to play a role as a public intellectual and to actually help develop understanding and consciousness about these issues. He understands, in short, the social responsibility of a progressive intellectual. And he, also, is not concerned with self. He also is willing to take some risks and to stand up to slanders and be demeaned for some of this. He refuses to go along—and he doesn't. I think one of the things that both these people show is a willingness to stand up under fire of a certain kind. We can talk about this later, but there are all sorts of people who wanna tear down people who are trying to change things in a positive direction.

So, not to go into that right now so much, but I just wanna say that there was something—I'm trying to find the words to describe the magical atmosphere. Here's the thing: I think there were some people who were in attendance...I heard that they said afterwards, I wish every day could be like today. And I felt that myself. It felt like you were in the presence of...that there was **leadership** in the room, that there was a diversity of people in the audience, that there was a shared concern about a lot of the outrages and injustices in society and a shared lively determination to do something about it, rather than just accept it as the way of the world. So it was very encouraging.

And there were many other things. I mean, even the venue. OK, look, I'm an atheist, I'm not a religious person. I don't believe in supernatural forces of any kind. I'm a scientist who is deeply steeped in historical materialism, and I don't get wowed or awed by the sanctity of religious places or religious venues. But that doesn't mean that I can't appreciate the beauty of the religious art. This church, Riverside Church, is a beautiful venue, and it has all sorts of interesting and beautiful carvings in wood and in stone, and so on. It's just a beautiful place, and you'd have to be stone cold not to be able to appreciate the art, even if you're not a believer. And this was a wonderful setting for this historic event. It is a church that historically has hosted many controversial subjects and topics over the years and has provided a platform for the contestation of ideas. And I thought that, once again, this happened in this period in a way that hasn't been seen in a long time, has never been seen actually. I can't think of another example of exactly this kind of event in history, where a revolutionary communist leader of the revolution is meeting together with a revolutionary Christian so that they can bring forward what they have in common and explore the differences and put it before the people and encourage hundreds, thousands and ultimately millions of people to engage these very important questions that have to do with morality and conscience and with the future of humanity.

And the topic itself is so important, the topic about religion and revolution. Look, I've been arguing, and I know this is definitely BA's framework, that you have to have a scientific materialist approach to analyzing the patterns of society—past, present and future—in order to figure out what to do about the problems of the world. Other people think that you have to apply a religious spiritual framework. That's a different approach to trying to deal with some of the same problems. It's a different approach to some of the solutions. It's a different approach, but it doesn't have to be in all cases an antagonistic difference. In this case, one of the things that I saw, and was inspired by, was that I thought there was a strategic alliance being modeled between a revolutionary communist...the revolutionary communist project, and progressive moral religious people, as embodied by Cornel West. Many religious people are not so generous of spirit, so moral, so solid in terms of their conscience. But this is an example of how two people can walk together and two actual sections of society can walk together in a strategic alliance. I thought that was very inspiring and should give hope to people.

And there was a lot that was modeled methodologically by BA in this Dialogue, in how he dealt with a lot of these questions. Many people are afraid to criticize religion—they think the people need it and you shouldn't say anything. One of the things I really like about BA is that he's never afraid to tell people what's what, even if he knows it will make them uncomfortable, even if he knows that it'll be controversial, that it's not the popular way of thinking, that he will be attacked or even slandered or reviled for doing so. He's just gonna tell people the way he sees things, on the basis of a scientific examination over decades of some of the key underlying phenomena. And, OK, religion, as Cornel expressed it very clearly, especially for Black people in this country, this is where many people live, this is very close to their heart, this is very intertwined with the history of resistance of Black people to oppression since the days of slavery, it's very intimately tied in with people's loved ones, and their feeling of who has led them in the past to fight against oppression. So it's all very intertwined. And BA is very clearly expressing to people that he understands all of that, but that you have to let a lot of this go, you have to let it go because it doesn't correspond to reality and it will actually take you off course and make it harder for you to actually transform the world in the direction that would benefit the majority of humanity. So there's a difference there, but it is a difference that can be wrangled with and analyzed and subjected to critical analysis and thinking. And the audience was into it. The vast majority of the audience was really into this—BA's presentation, Cornel's shorter but substantial remarks, and then the dialogue between them where they went back and forth. So there's a tremendous amount there. I think it's worth re-watching and re-viewing the live stream, and the upcoming film, because there's a lot to learn from what was being modeled there, and by the whole event.

So the speakers were great, the topic was great—and then I have to say about the audience: There was also a magical element, something that was greater than the sum of the parts, that came out of the connection, the presence of the audience with the speakers. That was something that I may be having trouble putting into words exactly, but I felt it very strongly at the time. There were 2,000 people or so filling this historic venue. And many came from the area, from New York, but many came from far away. There were people there from Chicago, from Ferguson, from Boston, from Hawaii, and so on. People actually traveled there, people raised money for some of their friends to be able to go and represent for them and for the ones who couldn't all go and travel such distances. So you had people arriving, you had buses arriving, there was an excitement in the air, you got the definite sense that people felt this is an important day, a day where we're going to talk about the things that are really wrong in the world, all the outrages and injustices, and, in particular, at that moment, there was a lot of focus on these police murders and brutality. And we're gonna talk about: do we have to take it, or can we put a stop to this, and how are we going to go forward from here? And partly it is taking a moral stand, but it's more than that. There was a lot of discussion with both speakers encouraging the people to stand up and fight this stuff. Both speakers were very good about doing that. And there was a certain electricity in the air when, for instance, the buses came and there were people from Ferguson who arrived, and they came in chanting, "Hands Up, Don't Shoot," and the entire audience...this was before the start of the program...the audience stood up and joined in: "Hands Up, Don't Shoot!" I'm getting goose bumps even thinking about that. And everybody felt it.

And part of what was really, really special about this was the **mix** of people. And this is something I give great credit to the Revolutionary Communist Party and the leadership of Bob Avakian for, historically, going way back to the '70s and since then. I don't know any other organization that brings people together in the way that BA's leadership and the Revolutionary Communist Party does, in terms of being able to bring together people from what are often referred to as basic masses, in other words, the people from the inner cities, the people who might not have much education, who are poor and the most oppressed of the oppressed, and for whom daily life is a constant struggle under the boot of the oppressors...bringing them together with students, college students and others, including older people, from the middle strata, from the intellectual strata, from the artists and the scientists, and so on. So you have a Ph.D. professor, or a prominent person in the arts, who is sitting with somebody who is from one of the hardest inner city ghettos in the country—and they're together! They're not looking at each other with suspicion. They're not looking at each other with fear or disdain. They're together in this because they are being brought together by this project and by this whole determination to put an end to this degrading and dehumanizing oppression, and to make a better world. And whenever I've seen glimpses of that, going way back even to the '70s, even in how I, myself came forward, that was one of the things that has inspired me.

BA talked at the Dialogue, very movingly, about Wayne Webb, also known as Clyde Young, and what a hard life he came out of, and how he developed and emerged as a leader who became a member of the Central Committee of the Revolutionary Communist Party, and what his whole life trajectory was about, coming from the hard streets and from the prisons. You have someone like that, and you have people who've gotten Ph.D.s in science or who are prominent artists or prominent members of society who can be in the same party and in the same movement for revolution. That tells you something. It doesn't tell you everything, but it tells you something important about the nature and characteristics, the type of movement that this is. And this bringing together—the great diversity of the audience being brought together, at this moment in time when people are waking up and standing up against some of these egregious police murders and other abuses in society, and becoming, once again, more determined to figure out if there's any real way we can change things for the better—and coming together with these two speakers who, in their own different ways, were speaking to the people. One was a speaker of conscience who describes himself as a revolutionary Christian who was encouraging intellectuals to have principle and integrity and to stand with the oppressed. There are not many people from the intellectual strata these days who are doing that, and I salute Cornel West for taking that position and promoting it and serving as an example of that.

And then you have Bob Avakian standing there, on the basis of decades of hard work developing a whole body of work—theory to advance the science of communism, to advance the science of revolution, to more deeply explain where the problems come from, what the strategy is for getting out of this mess, what the methods and approaches should be to stay on track and actually build a better world, to build a society that most human beings would want to live in. That's a hallmark of Bob Avakian's work, working on building a society that most human beings would want to live in. But, to do that, you have to understand the need to sweep away the system of capitalism-imperialism and to build a completely different society on a different foundation—economically, politically, culturally. He was bringing that to life. And he was also bringing to bear the strategy for today. You know, it was brief. [laughs] Some idiots were complaining that he spoke too long. Actually, a lot of people were glued to their seats and wanted to hear even more, if there'd only been more time. But luckily, we have his whole body of work and the website at revcom.us is full of books, articles, speeches. There's the film *Revolution—Nothing Less!* which is six hours of exposition from Bob Avakian's work, which people should really get into. There's *BAsics*, which is a really good book to start with, which also points people to the major works that things are taken from. So there's no shortage of materials to go to.

But at least, in that short period of time, you were able to get a feel for the strategy for an actual revolution, what it means to work towards that, what it means to provide leadership, what is the nature of leadership, what is the role of new people in relation to that, why everybody needs to come into this process, and there's a place for you no matter where you come from, there's a place for you in this process, in this revolutionary process, and there was a lot of modeling of the kind of culture in society that we would want to bring into being. And then there were some very serious discussions of the connection between the very necessary fights of today—the protests, for instance, around Ferguson, and so on—and the actual struggle for revolutionary power, the seizure of power. What is the connection, how does one help build the other?

And there was at least preliminary discussion of some of the work that's been done to bring out the real possibilities for how to actually win. That working on revolution isn't just a good moral thing to be doing—you actually have to do it in a way that you have a chance of winning and not being crushed. BA spoke about that some, and he pointed people to some key documents that are available on that revcom.us website: the documents "On the Strategy for Revolution"; and "On the Possibility of Revolution," which is a document that talks about the strategy for the actual seizure of power, and how you might have a chance of winning instead of being crushed by the forces of the other side. And then he was also pointing to the *Constitution for the New Socialist*

Republic in North America (Draft Proposal)—and...I have to say it just for a second here...that's an incredible document that I don't think enough people have actually looked at, or even leafed through briefly, to get a sense of it. There is actually a **Constitution**, for a new society, that has been developed based on the work brought forward by Bob Avakian, his whole new synthesis of communism. So if you wanna know what kind of society Bob Avakian's work is trying to bring into being and lead people towards, you have something very concrete that you can dig into, that talks about the rule of law under socialism and what kind of freedoms there would be, how you would organize the economy, education...I mean, every imaginable question.

So, there was a lot presented there at the Dialogue, in a short period of time. There was enough, I think, to whet a lot of people's appetite to actually go and dig further into this and join in, both in the struggles of today that are very necessary: again, things like the police murders, and on a number of other fronts, including what is happening in terms of women and the attacks on abortion—this is basically a way to enslave women, to deny them the right to their own reproduction, to control their own reproduction—and other abuses, and the wars, and the environment, and so on. There was an outlining of a lot of that, and then there was a pointing to where people could go to learn a lot more and to get into a lot more.

And something else I want to say about the Dialogue is that there was this wonderful, affectionate, warm, rapport between the two speakers, which was also a model. These are two people who, with their differences, care about each other greatly, and appreciate each other deeply. There was a lot of warmth, and both people just came off as really warm, generous-minded individuals, and there was just a wonderful comradely atmosphere between them that I thought also was in sharp contrast to the kind of culture that prevails today. It was a good model of how you could have differences—and neither one of them was going to throw away their principles, you know, they had their differences and they were going to make those differences clear—but not only did they also bring out all the points that they agreed on, and the need to fight injustice and oppression, but they served as a model of how to handle differences. This is very important: **They were modeling how people should relate to each other when**

they have differences. Because they were more concerned with the conditions of the oppressed and what to do about it than about themselves and their own egos. Because both people were much more focused on **that**, they found it in themselves to interact in a **principled** way, and with generosity of spirit and comradeliness. Nobody was—to be crude, nobody was kissing ass to anybody else. When there were differences, there were differences. But they were very respectful and principled and willing to dig into things. And that was **a model** that a lot of other people in society should actually be inspired by and try to emulate. This is what the people should do. When you have differences, you should struggle over **substance** and not...look, there's generally way too much of a culture in current society of nasty attacks and gossip and snarkiness and petty complaints and petty criticism. When people dedicate their whole lives, and this is certainly the case...you want to talk about Bob Avakian, he has spent his entire life dedicating himself to trying to serve the people, to trying to bring into being a better world, to fighting for **that**...he could have feathered his own nest, he could have just tried to make his own life better for himself. But this is not what he's done. He's dedicated his whole life to working on the problems of why there are so many outrages and oppression and so on, and what to do about it. That deserves respect, that deserves appreciation, and it deserves being looked into critically but deeply, to really try to grapple with what it is that he's bringing forward that is new and different and should be learned from.

Bob Avakian—A True Scientific Visionary

Question: One thing I wanted to zero in on a little bit on this point about what struck you in particular about BA—and I think you've definitely talked about some of that, but just to go a bit more at this point about BA's scientific method and leadership, which was in evidence during the Dialogue—I guess a way to put it is: For anyone who wants a fundamentally different world, or even people who are beginning to question why the world is the way it is and if it could be different, what lessons should people be drawing from the scientific method that BA was applying during the Dialogue and, obviously related to that, his leadership as it got expressed in that Dialogue?

AS: Well, if we're going to talk some more about scientific methods and leadership—using scientific methods and how BA actually concentrates that kind of scientific approach—we should be talking about truth and what truth is. Because I felt that this was modeled during the Dialogue. I believe BA quoted Malcolm X—and it's a quote I've always loved—I'm paraphrasing a little bit but at one point Malcolm X said something like, I didn't come here to tell you what you want to hear, I came here to tell you the truth, whether you want to hear it or not. I think that's pretty close to the exact quote. I love that quote, and I love the fact that BA embodies that same kind of approach and attitude. It's a very core part of his method. It makes his life more difficult, I'm quite sure, because it's always easier to pander to popular, fashionable views: what do people say, what do most people think, what do most people like or not like. A true visionary...I believe that Bob Avakian really is a true scientific visionary when it comes to the question of the transformation of human society, I think he's bringing in a lot that's new, he is building on the communist science and the development of communism through previous periods, but he's taking it a lot further and he's got some really important conceptions and methods that are putting the whole science of communism on a more sound foundation and a much more inspiring and hopeful foundation than at any time in the past. So I think there's a lot in his work to dig into.

And at the Dialogue, I felt that one of the things that came through is his commitment to truth. That might seem obvious in a leader—that, of course, you should be telling the truth—but it's not just that there are corrupt leaders who lie to people and manipulate the truth. Sure, we all know about that. But there are a lot of people, even well-intentioned people, who don't actually understand what the truth **is** in a scientific way. [laughs] There are actually people who function as if the truth is what most people think, or most people say. Well, if you stop to think about it for a minute, of course that's ridiculous, and Bob Avakian gave examples of that in the Dialogue, including in relation to religion. For instance, I remember the example he gave of epilepsy—that in times past and under the influence of old religions from thousands of years ago, when people didn't understand a lot of stuff, most people would have thought that epilepsy was caused by being possessed by the devil, and it's only in fairly recent modern history that people have understood that it's a disease and that it can be treated, and that it has nothing to do with devil possession or things like that.

But the point is that one of the things that BA consistently models, which is a hallmark of a good scientist, is **being willing to go where the evidence takes you**, and not looking at things superficially, but systematically and methodically digging into historical experience, and from many different directions—the historical experience of political forces, of revolutionary movements, of communist parties and movements, of the international situation—examining all that accumulated experience, and also drawing on other spheres, not just politics but also art and science and culture, all the many facets of human experience throughout history, in order to **draw out the key patterns and the key directions of things and the key contradictions** which come to characterize a phenomenon, or a particular phase of history, or a particular form of social organization. And then critically evaluating it, and figuring out **on what basis it could be changed** if it doesn't meet the needs of the people.

One of the things I'm struck by, as someone who was trained in the natural sciences, is how unscientific most people are! Even very, very educated people, people with Ph.D.s in different spheres or whatever, are generally incredibly unscientific. They just have knee-jerk reactions to things. Very often, very educated people come across, frankly, like blithering idiots when they try to analyze phenomena in society, and that's usually because they are basing themselves not on science but on populism, on what is the general consensus. I don't really care what most people think, if it's not right. You have to show me the evidence of why something is true. And if one person is putting forward something that is true (that corresponds to actual reality) and yet nobody else agrees with them, that doesn't make it not true! Show me the evidence. And, conversely, if great numbers of people believe something to be true—"everybody knows this" or "everybody knows that," there's a general consensus—I have to say that, as a scientist, I don't find that particularly convincing! You are really going to have to show me the evidence.

You can't just tell me the numbers, you can't play the numbers game, you can't tell me that something is true just because a lot of people believe it.

One of the things that really captures this from BA, and that can be found in the book *BAsics*, is the statement that I believe is a real concentrated expression of a scientific method on the question of exactly what we're talking about here: **What people think is part of objective reality, but objective reality is not determined by what people think.** That's worth pondering and reflecting on. That's the difference between subjective reactions to things and a real scientific method. Because what people think is important. It's either right or wrong, it should either be encouraged or discouraged, it should either be reinforced or transformed. But in any case it's part of objective reality and, so, of course, it's important. But objective reality is not **determined** by what people think, no matter how many people think it or how few people think it. You have to dig deeper, you have to dig and uncover those underlying features and patterns. And that's one of the things that is a hallmark of BA's work and of the new synthesis that he's brought forward. And it is in sharp contrast to what has too often prevailed in a lot of the political movements—even revolutionary movements, even communist movements—in past periods and even through today. It is shameful the degree to which there is not rigorous scientific pursuit of the truth among many people and many organizations. And it's a problem in the international movement, among international forces today. There is often an unwillingness to critically evaluate the past.

One of the things that BA has argued for a lot is that **we have to be willing to confront the truths that make us cringe**. If you're serious about trying to transform the world in a good direction, you have to be willing to examine past experience in a rigorous scientific manner. There are two parts to that: You have to dig deeply to understand what is correct in what was done before, in what was previously understood and what was previously accomplished; but then you also have to be willing to recognize where things went off track, or where there were shortcomings or mistakes made. That's how we learn, historically, that's how human beings accumulate knowledge, but it's also absolutely necessary for transforming things in the right direction.

And, you know, there are a lot of wrong tendencies epistemologically. Epistemology is the science of how you think about thinking, how you accumulate knowledge. That's what that is. And the question is, how do you know something is true? You should not be trying to determine what's true just on the basis of how many people believe it or don't believe it. You should also not be trying to say that the truth resides in superficial phenomena, like in an immediate narrow slice of experience or practice. You should not fall into pragmatism. Pragmatism is the view that if something works now, then it must be true. I was reading a good example about that in a very interesting piece that I would recommend people study. It can be found through the revcom.us website—it's in the online theoretical journal *Demarcations*, which can be accessed through the revcom.us website. In this piece, there is an important appreciation of Bob Avakian's new synthesis put out by the OCR, the revolutionary communists in Mexico, entitled "The New Synthesis of Communism and the Residues of the Past" by the Revolutionary Communist Organization (OCR), Mexico. It's about some of the line differences in the international communist movement, and it's an appreciation of Bob Avakian's new synthesis in relation to that. And there's a whole discussion of pragmatism in there, and how many people think that truth is whatever is kind of "convenient" for accomplishing certain objectives in a very narrow and immediate sense. The article gives the example of the thalidomide drug which was developed some time back to treat morning sickness and was touted as an advance in science. Well, it "worked" for that purpose and it got heralded, but it turned out that it hadn't been sufficiently, deeply analyzed in an all-sided way, and it also led to children being born with tremendous birth defects. The deeper truth turned out to be how harmful it was, not that it "worked." Well, that's an analogy for the same kind of mistakes that can be made in the political sphere.

And Bob Avakian insists that everybody should act like critical thinkers, and really that everybody should contribute to the process of actually analyzing what is true and what is false in various kinds of phenomena. It doesn't matter who you are, how much experience you have—you can be in the Party as a Party leader, or you can be in the Party as a new person and relatively inexperienced, or you can be outside the Party, you could be a critic of communism or you could be an adherent of communism—it doesn't matter who you are. If you have principled methods, and you are willing to actually try to get to the truth of things, your contributions would be welcomed in terms of trying to advance knowledge and understanding. Now, you also should be willing to be subject to criticism yourself, from others who might punch holes in your theories or analyses. That's what good scientists do. As a natural scientist, I had many good experiences that way, where I or other scientists would put forward some analyses of some things in nature and propose some experiments that could be conducted to uncover some of the deeper reality, and then you got your colleagues and friends together and they would spend the next hour or so trying to punch holes in your theories and questioning your underlying assumptions! That can be a very healthy and productive process (and fun too!), as long as it's done in the right spirit (free of snark and ego) and with the right method.

The New Synthesis of Communism, Solid Core and Elasticity

Question: Well, we're definitely gonna get more directly into the new synthesis of communism that BA has brought forward that you mentioned, pretty soon. But in terms of the method that BA models in all of his work, including at this Dialogue, one thing that was called to mind for me by what you were just saying is the relationship between the point you're making about constantly going for the truth and another key dimension of this new synthesis of communism, which is the approach of solid core with a lot of elasticity. So, I wondered if you wanted to also talk about how this Dialogue was an example of applying solid core with a lot of elasticity.

AS: Well, I think this relationship between solid core, and lots of elasticity on the basis of that solid core, is a real hallmark of Bob Avakian's entire body of work, of the whole new synthesis. It is in evidence in everything he does and writes and talks about, and it was in evidence at the Dialogue. I mean, one of the things you can count on with Bob Avakian is that he will tell you what his most developed and advanced analysis and synthesis has brought him to understand. He will share that with the world without hesitation, regardless of how popular or unpopular it is, and he will back it up with evidence. And anybody who wants the proof can look into his works and how he got to certain things, certain conclusions about the nature of the system and the way forward, and so on. But one of the things that he's understood, in the course of studying deeply the experience of the first wave of socialist revolutions, and the positive and negative experiences of the past, is that he's come to appreciate even more deeply the need for a scientific method that you might think of as neither too rigid nor too loose. [laughs]

Good science, you know, does not just go out into the world with a big question mark without any kind of developed theory. In order to advance science, you go out into the world with a framework of certain analyses that have accumulated over time. You make your best possible analysis and synthesis at any given time. And then you go out and test it further against reality. That's what scientists do. And, in the course of that, you discover that some things that you thought were true are in fact very much true—you see some patterns that maybe you expected—and you often also get some surprises, you learn some things you didn't expect, you find out you were wrong in some instances, and you learn from that as well. That enables you to make an even more advanced analysis and synthesis. And you go on from there. That's how good scientific knowledge advances. And Bob Avakian models that in everything he does, in my opinion. That's why I think **there's really no one like him in terms of taking a really consistently good scientific approach to societal issues and the positive transformation of society**.

And what you saw at the time of the Dialogue—you're looking at a guy who's really **a statesman**. People say sometimes, "Well, we need to change things, but there's no leadership." Well, you want leadership?—**there's** leadership. There is leadership that is not hesitating. There is leadership that has the confidence that has been built up on analyses and syntheses of world experience, and the experience of this country, and the experience of the communist movement and revolutionary movements, on the whole human history of experience, which has been studied and analyzed for decades. He's got a lot under his belt that way, and he has no hesitation at sharing with the people what he has learned, in a very coherent way. **That** is leadership, and that is **the solid core** of his leadership.

At the same time, it's very much part of his scientific understanding of things that communism and the transformation of society—this is not a religion, this is not a dogma, this is not catechism, it is not a set of precepts or rules, it's not the Ten Commandments. It's a living science that must always be open to learning from some new directions and new experience and new information, new data coming in, which can both reinforce and further substantiate what you already understand, and also call some parts of it into question and allow you to develop it even further. It's not a static process. Science is a very dynamic process, correctly understood. So, one of the things that you see is...why is he even bothering to do something like this Dialogue? Why is he speaking to such a diverse audience? It's not like most people there were communists. Most people there were not won over to everything—he's not preaching to the converted. Again, **it's not a religion. He's bringing science to the people**, and he's calling on people to engage it and to bring some of their own experiences to bear and bring new insights into further deepening the truth and further deepening analyses.

But what he's also not doing is making the opposite mistake that people can make. On the one hand, there's **the mistake of dogmatism and a religious approach**—acting like, instead of a science, communism is just a bunch of precepts or a catechism that you should recite and that has that kind of rigidity. No. Real life, real nature, and real human society is much too dynamic to be forced into these dry little precepts and cubicles! **But the opposite mistake people can make epistemologically is to act as if nothing can ever really be known, nothing is ever certain.** Acting like, just because it's right to question everything you can never be sure of anything, that there's never anything you can ever base yourself on to go forward, to learn more—basically arguing for all elasticity all the time, so that there's no longer any solid core to anything at all. It's like what prevails a lot in university circles these days: a tremendous amount of philosophical **relativism**, where people will literally say to you things like: "Well, there's your truth and then there's my truth, we can all have our truths, and you can have your narrative and I can have my narrative, and who's to say what's right or wrong." In my view, such extreme relativism is not just idiotic, it's unconscionable.

If you never have any scientific certitude about anything, you're going to have a lot of disasters and you're not going to move forward. For instance, in the natural sciences, if you're trying to solve a huge environmental problem, or cure a serious disease, or send a probe into space to explore Mars, or whatever, you'd better be starting off with a certain solid core of scientific certitude, to the best of your ability, even while recognizing that some parts of your understanding and approach may not be perfect. In fact, you can almost always predict that you will be learning some new things that will call some parts of your understanding and approach into question. But you'd better start off with an initial scaffolding or template which involves a certain core of certitude, of scientific certitude, that has been built up over time through the accumulation of historical experience and the subsequent scientific sifting through and "triaging" of that historical experience. This allows you to say, OK, we're going out into the world applying these scientific hypotheses and theories, and we're going to further test them and develop them and no doubt learn many new things along the way. But if you don't set out with some scientific certitude, with some solid core to put out in the world, you won't be able to accomplish anything. If you don't think there's anything you can reliably base yourself on...you might as well be floating in a vacuum! If you're trying to figure out how to cure cancer or some other terrible disease, you can assume that there will likely be some flaws and shortcomings in your understanding at any given time, but you'd better be willing to apply the best accumulated understanding to date and use this as a basis to further experiment and try to transform reality, and then to further sum up and analyze in order to enable even further advances in solving the problems.

Bob Avakian models that kind of scientific approach. He doesn't tell you everything in the past was perfect. Or that everything in his own understanding was perfect. Or there won't be any mistakes made in the future. He never tells you that. He tells you that we have to learn to apply a scientific method and approach in order to systematically analyze and sort out what is true from what is not true to the best of our ability at any given point in history and in an ongoing way.

And, of course, if you're talking about social transformation, you have to attach your scientific method to a moral conscience. You have to actually be proceeding back from certain objectives. If you're a natural scientist, maybe your objective is to figure out the effects of deforestation of a rain forest, or something like that. If you're a social scientist and you're a revolutionary communist, your objective is to actually move towards a better world, a world that transcends class divisions and accomplishes what's been called the "4 Alls." The "4 Alls" refers to a formulation by Marx where he said that reaching the goal of communism requires the abolition of all class divisions, of all the production relations on which those class divisions rest, of all the social relations that correspond to those production relations, and the revolutionizing of all the ideas that correspond to those social relations. You're moving in that direction of a communist society, and you're understanding some of the contradictions involved today in the process of seizing power, in the process of building a new socialist society on a completely different economic footing and with different objectives and different social relations being fostered and fought for. **You're moving in a certain direction.**

So, today a revolutionary communist should be proceeding back from that longer term goal of bringing into being the kind of society that would be truly emancipatory for the majority of humanity. And that should be what you're continually double-checking: Is the work going in the right direction? Is it moving towards, rather than away from, those stated objectives? The work may go down some wrong paths, and you can hopefully recognize that soon enough and correct course and learn from those mistakes. Any good scientist will tell you that you can even learn a lot from your mistakes and your wrong directions, as long as you consistently apply scientific methods to their analysis and summation. But if you don't apply consistent scientific methods, you are much more likely to get devastated by mistakes and misdirections.

Bob Avakian has said that all truths are good for the proletariat, that everything that is actually true can help us go in the direction of communism. And that's really true. You can learn. And in the Dialogue you can see him actually struggling with the audience to understand that. He knows he's talking to an audience that harbors a lot of misconceptions, a lot of prejudices—it's an audience that is full of intelligent people but people who do not know much of anything about how society is structured and organized, or what it would take to actually remake society on a much more positive basis. There's very little science, there's very little materialism in society today. I don't care how educated people are, most people don't know anything, frankly, when it comes to understanding society and how to transform it. I don't hesitate to say that. And Bob Avakian is modeling the solid core, in that he's saying, Look, I've been at this for decades, I've been applying scientific methods. There's a lot that I've learned, there's a lot I can share with you about how this system is constructed, about these outrages like police murders and all these other outrages. Why do they happen? Why do they **keep** happening? Why will they continue to happen until we get rid of this system?

He has a lot of solid core material, a lot of scientific certitude that he can bring to bear. And, at the same time, he has these very wide arms, that are open to bringing together people who have a lot of different perspectives and to drawing from broad insights and experience, including, in this case, his very warm and productive rapport with Cornel West, who proceeds from a different philosophical epistemology, but who shares many of the same concerns.

There is much we can learn from these diverse frameworks, but they have to be sorted out and they have to be harnessed and directed. The elasticity shouldn't just be a random mush. It should constantly be brought back together with the solid core to direct it and channel it. A good scientist does try to direct and channel things in positive directions in order to resolve problems. And that is part of what you're seeing—you're seeing it in the course of the Dialogue. There's a great deal of confidence and certitude of the scientist who's done a lot of work, and who knows that his work is very advanced, and who knows that many of his critics have never really engaged the material with any real substance.

And, at the same time, he's opening his arms out very widely, both in encouraging critical thinking and in learning from a lot of different experiences in the past and in the present, all for the purpose of leading in a direction that would actually be good for the majority of humanity.

Another thing you could get from the Dialogue, in terms of what Bob Avakian was modeling, was something of a feel for the kind of society he's arguing should be brought into being. I think people are often surprised when they read or in other ways encounter Bob Avakian. People often come with all sorts of societal prejudices, misconceptions and stereotypes, about communists being some kind of dry and humorless dogmatists, but then they encounter Bob Avakian and discover that he is completely different from what they expected. And this is precisely because of the kind of scientific method and approach he takes to the transformation of society and working towards the goal of emancipating all of humanity. He's very lively, and he has a tremendous generosity of spirit. And he's very funny. That's always something you hear people say—I never knew he was so funny! At the same time, he's dead serious, he's absolutely dead serious about what he's about. And his rage, his anger, his outrage at the injustices of society, the depth with which he feels every one of those police murders of Black youth, for instance, he's not putting on a show, this is something that is profoundly felt. Everybody comments on that—that his sense of outrage is very real, his seriousness and determination to do away with all this is very real. And, at the same time, he can combine that hard core seriousness and science with an approach that is lively and generous and full of humor and that embraces life in all its many dimensions. And that, I think, says something about the kind of world that he's arguing to bring into being, and the methods for doing so.

A Communist Statesman, Modeling Communist Leadership

Question: I think that that's a really important point, and it relates to something you said a minute ago, that you felt BA really came across as a statesman in this Dialogue. And maybe you could explain that a little bit more, because I think that's a really important point and I know you were saying earlier that you felt like you really got a sense being at this Dialogue, experiencing it, that this is the leader of the revolution, this is somebody who could lead the future society. So I don't know if you wanted to speak a little more to that.

AS: Yes, the reason I felt the statesman aspect, too, is that I think we live in a complicated period, that there are a lot of challenges in this period to actually advance the revolutionary struggle, to deal with the actual fight—the "fight the power, and transform the people, for revolution" aspect of things is going on right now in a way it hasn't for some time, in particular around the police murders. And, look, BA leads the work of the Revolutionary Communist Party, and there's not a single initiative, I'm sure, of the Revolutionary Communist Party, that doesn't have the stamp of leadership of BA and of the top leadership of the Party on it, in terms of how it's unfolded. As you can see from the diversity of things that are taken up by this Party, and as reflected in the website revcom.us, there are a lot of very challenging contradictions to deal with. And that gives only a hint of what this leadership involves.

I don't think most people have any idea what revolutionary leadership is about. A lot of people think that a leader of revolution is kind of like an "activist" leader, sort of like a leader of a demonstration, what I think of as **tactical** leadership. But overall revolutionary leadership is not just tactical. Of course there does need to be tactical leadership in various dimensions, and I'm not trying to devalue that. There is very much a need for the kind of person who might be agitating in a demonstration, for instance, helping to put forward a better understanding of what people are fighting about, and leading people, even tactically, in the streets, for instance in a demonstration. But there's an important point to be made about how the leader of a revolution and the leader of a new society has to be **an all-round statesman** and has to be more like **a strategic commander** of the revolution as a whole. And there's a formulation that's been put forward recently that a communist leader—and not just the top leadership, but every single revolutionary communist—has to think of themselves and strive to be a strategic leader of the revolution, "a strategic commander of the revolution, not just a tactical leader, and not just a strategic philosopher." This is very important. In other words, if you're going to lead a revolution, lead the seizure of state power and become a leader of a new society—and that's what I mean by statesman—you have to fully recognize and grapple with the complexity of what you're doing and the many different levels and layers of it, and the many different contradictions among the people. You have to deal with the fact that you don't have absolute freedom at any given time, and yet you're trying to move things in a certain direction. You're trying to be true to your principles, and you're promoting that openly, but at the same time you're dealing with the people you're leading, who often don't understand, at least not with any depth, what you're putting forward in leading them, or who tend to distort what you're putting forward, because they don't understand things well enough or because they're being shaped and influenced by other programs, other outlooks and methods.

So strategic leadership is a very, very complex task, and that's also involved in why, as I mentioned earlier, so many natural scientists are at a complete loss when they try to address social transformations and they suddenly seem to forget everything they ever knew about basic scientific methods! Part of that is also because so many people have a completely wrong view of what actually constitutes overall leadership in the social arena, especially as pertains to revolutionary change. Much of the time they seem to think a political leader is just somebody with a bullhorn in a demonstration. But that's tactical leadership, that's not the overall strategic commander type of leadership that can guide an actual overall radical transformation of a whole society through revolution and the building of a whole new kind of society on a fundamentally different economic basis, with everything that flows from that. That kind of multi-faceted leadership is a much more complex task, and most people today frankly have little or no conception of all that it involves.

And there's the question of dealing with the audiences—if you wanna put it that way, there are many different audiences. You're not trying to be all things to all people. You are actually trying to meet the objective interests of the international proletariat, by which I mean—it's not any individual proletarian that's the question—there's an international, world-wide class of people who don't own the means of production, who have no ability to run society under this system, who can really only sell themselves basically, under this capitalist-imperialist system. They have the greatest interest—whether they know it or not as individual proletarians—as a class, they have the greatest interest of any class in actually going in the direction of communism and getting beyond all these class divisions and relations of exploitation and oppression. But is that the only class that's going to be part of the process? No. The capitalist-imperialist ruling class is a very small segment of world society, or of any given society, but you do have all these other forces that kind of have one foot in one system, while one foot may be aspiring to something better. And those more "intermediate" strata, they tend to not be very constant, they tend to flip from one side to the other on any given day! Add to that the fact that hardly anybody has been given any scientific training, so hardly anybody tries to approach problems with any kind of consistently systematic and rigorous method. So you've got people going all over the place, you know, both in their thinking and in their actions. Bob Avakian's talked about the challenge of "going to the brink of being drawn and quartered," both in terms of getting to the revolutionary seizure of state power, and in terms of building a new society—that there are so many different kinds of people pulling in different directions, with different and opposing ideas, and so on.

And here's another reason you need science. How can you know what's best for society? **How can you know what's best for the majority of humanity?** The capitalist-imperialists, they are proceeding on the basis of what's best **for their system**. It's not just a question of corporate greed, it's not just that. It's much more than that. They have a system that they need to maintain, a system that is based on profit, and we can talk about the fundamental contradiction of capitalism-imperialism, it might be worth touching on that a little bit. But the point is that they're trying to keep their system going, but they don't understand—even the people running this society often don't even understand the deeper laws of their own system. But if you're trying to bring into being a whole new kind of society, one that actually more fully meets the objective interests and needs of the vast majority of humanity, you've gotta do a lot of work, and you've gotta go up against a lot of misconceptions and prejudice and anti-scientific views. You have to deal with that diversity of views and opinions and with people pulling in all sorts of different directions, while at the same time not losing the reins of the process itself. That's where the strategic commander role comes in. If you are confident in your scientific approach, then you can say with quite a bit of certitude that you think it is possible to determine what is in fact **in the objective interests of the majority of humanity**, and what it would take to move in that direction. It's like if you're riding a horse. You've got your hands on the reins, so you're not just going to let the horse run to any old place—the horse here being the process, not the people, but the process, right, the revolutionary process. But if you ride a horse and you pull the reins in too tightly, and you pull the horse's head too hard, and the bit cuts into the horse's mouth, and you're not allowing it any kind of free rein, then that horse is going to stop dead in its tracks, or it's going to buck, and in any case it's not going to be able to be part of freely moving forward and advancing the process.

So there's always a tension—the reason there's a need, as BA has stressed, for "lots of elasticity, on the basis of the solid core" is not, as some people have incorrectly argued, just because the middle strata of people are going to "buck" and cause problems for you, are going be resentful, and so you'll have to give 'em a bone here or there, to keep 'em from fighting you, or something. No! That would be disgusting. The real reason that you need to build in and allow for some genuine elasticity, on the basis of the solid core, is because society needs it, the process needs it. The revolutionary process itself needs to breathe, the revolutionary society needs to breathe, or it won't be any good. Both the process of getting to the revolutionary seizure of power, and then the process of building the new society needs to breathe. And if you try to control it all too tightly and too rigidly—even if you happen to be right in what you're doing at any given time, if you're too tight and controlling, it's just going to be discouraging and demoralizing to people, and people are not going to be given the scientific tools to figure it out enough themselves, and you're going to end up with a repressive society, a rigid society and a rigid process.

And Bob Avakian really understands that, because he's a good enough scientist to understand the material tension that exists, objectively, between what's called the solid core, the certitude, the elements that you can actually be confident of, in terms of what's wrong with the current society and what's needed in a future society to benefit humanity, while at the same time understanding the need to sort of shepherd the process in such a way that it can encompass and incorporate the widest possible diversity of views and approaches from among the different strata of the masses in society.

I don't know if I'm expressing this well enough, but he has certainly expressed this very well in many of his writings and talks, and I would encourage people to dig into this whole aspect of solid core with lots of elasticity on the basis of the solid core. And that last part—**on the basis of the solid core**—is very important to understand. **You couldn't have the right kind of elasticity without the solid core.** You don't wanna end up like you're trying to herd cats, with everything and everybody going all over the place. There does need to be a solid core. In fact, the more you've got a firm handle, a rigorous scientific handle, on that solid core, on that core scientific theory, on that core accumulated knowledge and experience and on that core certitude, the more it should actually be possible to unleash and encourage broad elasticity and initiative among the people, both in the current revolutionary process as well as in the future socialist society, including in relation to the kind of dissent and broad societal ferment which can actively contribute to further advancing society in a good direction.

Question: As you were talking, one thing that is posed is that there is a unity, there is a connection between what you're saying about the approach of solid core with a lot of elasticity, both in the process of making revolution to get to a future society on the road to communism, and then in that future society itself—there's a connection between that approach all the way through the process of making revolution and getting to communism and your point about how you could really get a sense in this Dialogue of BA as the leader of that future society. And then there's the point that you were making earlier, about why would BA do this Dialogue with Cornel West, if he weren't actually applying and modeling that approach of solid core with a lot of elasticity? And so something I wanted to probe a little further is this point about how BA, in this Dialogue and in his whole body of work, he's very much pulling no punches, he's very much putting forward his understanding of the science of communism and of reality, and he's not trying to finesse or smooth over differences, including with Cornel West, while at the same time he's also very much recognizing the unity that they have, and the unity that needs to be forged broadly. And he's taking the approach that there's a lot that somebody like Cornel West—he has a lot of insights, there's a lot that he can contribute to this whole revolutionary process, even while they're very much getting into their differences. So, is there more you wanted to say about the application of solid core with a lot of elasticity even in terms of how BA was relating to Cornel West in this Dialogue?

AS: Well, I think you can see the application and modeling of "solid core with lots of elasticity on the basis of the solid core" in what BA does, both in relation to Cornel West on the one hand, and also what I was trying to say before in relation to the audience—or audiences, plural, because there are many different strata and different viewpoints represented in the audience—and what you see is, you see the certitude based on experience and knowledge. Look, think about in the natural sciences: If somebody happens to emerge who is the most advanced in their field of science, or in a particular development of the natural sciences, at a given time—somebody who is really advanced and really visionary and really is playing a leading role that way—it would be ridiculous for them to come out and just kind of act as if they don't know what they know, or not struggle with people and not provide the evidence that they've accumulated and analyzed over, literally in this case, decades. Right? So even as he's working **with** Cornel, he's also not pulling any punches because, first of all, he respects people enough not to pander or condescend or pretend he doesn't know what he actually knows. The only people he doesn't respect are the exploiters and oppressors at the top of society. But he has enough respect for people, even people who might disagree with him in some important ways, to be honest and to explore differences with principle and integrity instead of condescending or pandering to people or pretending to have more agreement than he does.

He's gonna call it like it is. He's gonna tell people, including the audience...he knows this audience is holding on to a lot of different views and misconceptions that he thinks are very harmful. Like a lot of these religious views that are holding people back from understanding reality the way it actually is, and from seeing how it could be changed. His position is definitely not neutral—with religion, he's not just saying look, that's not where I'm at, but it's all good, go ahead and believe whatever you're gonna believe. He's definitely not saying that. Instead, he's really **struggling with the audience**, right down on the ground—he's saying, you gotta give up some of this religion stuff, because it is actually harmful; it is clouding your understanding of the way reality really is; and, because it's doing that, it's actually making it harder for you to see the way forward, and to see how to transform society in a good direction. So you gotta get off this stuff! And he's saying that to an audience of people, **most** of whom are religious, especially among the most oppressed—the very people who are most important for, and who most need to step forward to take up, the revolutionary process. He's got enough respect, enough strategic confidence in people, to tell it like it is.

Now, in the situation where he's working with Cornel, he's working with a developed intellectual who's also got a lot of experience in life, and who has studied many different things himself and analyzed many different philosophies. And BA's got respect for that process, too. But he's still going to call it like it is, and he's going to bring out the evidence. What does it actually <u>say</u> in the Bible? What <u>is</u> the role of religion? Let's get into it!

Some people might say, Well, I don't need to hear all this, because I already don't believe in God. Well, yes, you **do** need to hear all this, and do you know why? Because **billions of people** around the planet are deeply influenced by one or another religion, and they approach all of reality through the prism, through the lens, of their particular religion. This is the framework, this is the theoretical framework, if you want to call it that, that **most** people on this planet apply to try to make sense of the world, and of what's wrong with it, and what could or couldn't be done about it. Religion is a very major question, in the United States and all over the world. So Bob Avakian, on the one hand, in the Dialogue, you see him struggling with Cornel, but with a good method, a good warm method, because these are two people who do respect each other and who do like each other but who are just going to honestly tell each other and the audiences where they have some significant differences. And because they have principle and integrity, they're able to put forward and clarify those important differences, so that the audiences will be better able to grapple with these questions themselves, when they go home and in an ongoing way.

At the same time, what I think Bob Avakian is modeling, with the elasticity part, is: Listen, this revolutionary process, it's a very rich and complex and diverse process, **which does have to involve a wide variety of people**. In fact one of the points Bob Avakian has made repeatedly is that, at the time of the revolution and the actual seizure of state power, most of the people involved in the revolution are **still** going to be religious! In a country like the U.S., there's no question that this is true. Most people won't have given up their religion—even if they've decided to join in to be part of fighting for revolution and for socialism in different ways, most still won't have completely broken with all that. And that's just one example of having a materialist scientific understanding of reality, understanding just how complex it is, how complex the process is. But you're not going to try to trick people who disagree with you into walking alongside you in the revolutionary process by concealing your views. No, that's not what you should do. Instead, as a revolutionary communist, you're going to be honest about those differences. But, if you're serious about wanting to transform society in the interest of humanity, you're also going to recognize that the process that you are arguing for, and that you are helping to give strategic leadership to, has to be able to encompass quite a diversity of people, who are not all going to see eye-to-eye with you on a number of different and important questions. And that this will be the case all along the way, even as people increasingly unite together to fight the common enemy, to seize power, and to build the new institutions and organs of a new society.

It's because he really understands all this that Bob Avakian can, at one and the same time, genuinely and sincerely embrace and feel very warm towards someone like Cornel West (and I believe those feelings are very much reciprocated), and at the same time remain very clear about the importance of speaking to the differences, and speaking to why you need to take up a consistently scientific method and approach if you really want to change society for the better. And so yes, he'll tell people bluntly why they should give up religion—all religions—because they get in the way of moving forward. It is a fact that all religions all around the world were invented long ago by human beings, to try to explain what they didn't yet understand and to try to meet needs that can be transcended now. All around the world people invented different sets of supernatural beliefs to try to fill gaps in their understanding of things, in both the natural and social world, and as a mechanism for dealing with such things as death and loss. If you don't yet have the scientific knowledge to understand how all life evolves, and how there is clear evidence that human beings themselves simply evolved from a long series of pre-existing species, you're probably going to want to involve some kind of higher supernatural power to explain how we got here! [laughs] Every religion in the world has some of those commonalities. At the same time, they all have their different particular creation myths, and so on. And they have their different holy books, and prophets and stuff like that. And Bob Avakian is saying, Come on now, let's get serious, let's actually open up the Bible and see what it says. See, a dogmatic revolutionary might have said, Well, I don't believe in god, and I think religion's bad for the people, so I'm not even gonna pay any attention to it. But instead BA's saying, religion's a very important problem in the world, it's a very important question, billions of people believe in some kind of god or some kind of religion, so we have to address this. And he did some homework, too. He did the work. He read the Bible, in its entirety. He **knows** the Bible. Unlike many people, he can tell you what's in it. And he can tell you what these religious forces have argued. He can tell you something about the history of how human beings invented a lot of these religions. He can also speak to why people might be motivated to have a moral conscience on the basis of some of the things they learned in church or mosque or temple or whatever. At the same time, he can also show you, scientifically, the harm that it does to cling to this. And that it is not necessary. You can leave that stuff alone. You can just let it go. You can leave those old ways of thinking behind, and you can take up a philosophy and scientific method about transforming the world in the interests of all humanity, which is full of life, full of joy, full of spirit, full of art and culture, and not dead and cold in any way, but that doesn't have to have these religious and supernatural trappings and all the old stuff that goes along with it.

A Living Refutation of Stereotypes and Misconceptions

Question: What you were just saying brings up another point I wanted to raise. Earlier, we were talking about the stereotypes that people have about science—that people very often think of science in general as being very dry or rigid or lifeless—and it occurs to me that people will often have the same kind of conceptions, or misconceptions, about communism and communists. So I wondered, just off of what you were saying, if you wanted to compare, or contrast, those notions of science and of communism with the way that BA and his work and leadership comes across, in the Dialogue and more generally in his body of work?

AS: Well, that's an important question. I think one way to look at it—we've talked some already about the prejudices against science and how in particular there's a lot of **anti-science** being promoted in this society, and it's all the better to fool the people with. You can see that all the time. You can see it, for instance, in the political and ideological battles around evolution in recent years. The scientific reality is that all life on this planet (including people) evolved, over hundreds of millions of years, and none of it was specially created by gods or any other supernatural forces. That basic scientific fact has actually been very clear since the late 19th century, since the days of Darwin, and our understanding of this has only deepened and been verified 20 million different times from 20 million different angles in the more than 150 years since then. In short, there's an incredible amount of concrete scientific evidence that has accumulated over a very long time now and which leaves absolutely no doubt as to the reality of the theory of evolution—it has been **proven**, over and over again, beyond a shadow of a doubt: All life on this planet has evolved, and continues to evolve, through entirely natural biological mechanisms. Human beings themselves are simply one product of that long and diversified process of natural biological evolution. And the entirely natural basic biological mechanisms which drive evolutionary change are also well understood. Despite all the well-established scientific evidence, the people who run this society do not actually take it upon themselves to promote scientific education and the understanding of evolution in any kind of consistent way. They allow a lot of bullshit to be promoted to confuse people, foster crass ignorance, call biological evolution into question and promote instead religious belief in supernatural forces. Why do they do this? One obvious reason is that exploiters and oppressors can always benefit from the masses of people being kept in a state of profound ignorance and confusion. You see it in relation to science in general. The people running society could promote a view of science and scientific exploration that is lively and full of passion. And they actually do that to some extent, but generally this is only for a small sliver of the people, those they hope to train to be in their elites. But more broadly in society they allow an atmosphere to be fostered in which many, many people are confused about the most basic scientific facts and wrongly think that the process of science is inherently dangerous or deadly or dry and devoid of passion. I was saying earlier that science is actually full of the spirit of curiosity and adventure that typifies little kids and is an expression of the irrepressible human desire to better understand nature and society, the world around us, to make sense of it, to understand how things work and how they came to be a certain way and how they're changing. And there are a million questions on a daily basis that make science very lively and fun, actually fun, and a great method to apply to all aspects of life. And no matter what your circumstances and conditions of life are, or have been, you can definitely learn to be scientific. And it's a very liberating feeling to take up the basic scientific methods even in your everyday thinking and everyday life, as well as around more strategic questions involved in how to radically change society for the better.

But I guess the people currently running society would prefer you not know this! [laughs]

There are a lot of similar prejudices and misconceptions about communists and communist leadership that the people running things are only too happy to spread and promote. They promote stereotypes of communists as dry and dogmatic, bossy, scary, and they tell everyone that communists would repress any kind of individual expression and take away all your liberties. And they spread the notion that you better watch out for them, because they'll probably lock you up or something, or put you up against the firing wall. I mean, some of this reefer-madness type depiction of communists, it's frankly because the people who run society, the capitalist-imperialists, recognize on some level that the philosophy and ideology that threatens them and their system most profoundly is actually communism. And this has been true since the very beginnings of the science of communism, beginning with Marx in the late 19th century. The science of communism analyzes the deep roots of what's wrong with their system and how their system cannot be anything other than a nightmare for the vast majority of people and why it is that humanity cannot be fully emancipated without thoroughly dismantling the existing system (capitalism-imperialism) in order to bring into being a new socialist society and eventually move beyond that to full-out communism. And the science of communism comes with a method and approach that can help us figure out how to get there. The people running things don't want you to know any of this. [laughs] So they've **always** worked to slander communists and distort what they're about—and never more so than since the reversal of the most important socialist revolutions, in the Soviet Union and in China, where socialism has been reversed. The socialist revolution was reversed in the 1950s in the Soviet Union and in the late 1970s, after the death of Mao, in China, and it is important to understand that capitalism-imperialism has unfortunately been restored in both those countries.

In fact it's important to know that there are no actual socialist countries in the world today. There are no genuine socialist countries. Nowhere are communists, real communists, in power anywhere in the world. This is going to have to change.

There was, some time ago, what has been called "the first wave" of socialist revolutions that made great progress for a time. They also made some serious mistakes, which Bob Avakian has also spoken about. But it is undeniable that, for a time, they made great progress. Unfortunately today, even if some of the people running countries like China, or Cuba, still use words like socialist or revolution or communism, those societies, and those leaders, have nothing in common with genuine communists. People really need to learn scientific methods, including so as to be able to tell the difference between phony communists and real communists, and the difference between a real communist movement and a phony communist movement.

Part of what's happened, though, is that in the last few decades, since the reversal of the Chinese revolution in particular, the people running this society have waged systematic campaigns of slander to spread crass lies, claiming for instance that Mao killed tens of millions of people, or, you know, frequently lumping Hitler, Stalin and Mao together to make it seem like they were all the same and to denounce them as world class monsters. They actually have the nerve to compare communist leaders to a Nazi leader like Hitler! This is totally outrageous! But they love repeating that mantra, "Hitler-Stalin-Mao," to confuse people and make it seem like communists are just as bad as Nazis, bogeyman monsters crawling out from under your bed to stifle all individual initiative, lock people up and execute them. All these scenarios, comparing communists to Nazis are crude propaganda and total bullshit! Hitler and the Nazis were, of course, monstrous, murderous **fascist** oppressors. But communists, real communists, are the exact **opposite** of such fascists.

As part of their systematic campaigns of anti-communist slanders, the people who rule this society, and who dominate and oppress so many people here and around the world, have also gone to great lengths to encourage and promote the publication of novels and stories written by people who had belonged to the more privileged and parasitic strata in China, or who had family members who were leaders of the revolution and expected that this should bring them special privileges in the new society, and who felt resentful about some of the restrictions imposed on them in socialist China—restrictions largely aimed at limiting the ability of people like themselves to lord it over other people, especially from the more basic strata, as had been the case for centuries in pre-revolutionary China. Some of these people or their relatives have since written bitter and resentful "sob story narratives," to complain about the restriction of such privileges and to whine about how, boo-hoo, they "suffered so much" when Mao was still alive and China was still socialist. The widespread promotion of these subjective sob stories, in particular on college campuses, contributes to spreading really crass distortions of the actual (and definitely much more positive!) experience of socialism in China prior to the capitalist restoration, as experienced by literally hundreds of millions of ordinary people in China who supported Mao and socialism and the Cultural Revolution but whose stories the capitalist rulers obviously have no interest in collecting and disseminating.

I'd really like to see more critical thinkers on college campuses and everywhere else do some serious digging into established facts and examine the truth of that whole overall positive experience and stop getting taken in so easily by the cheap propaganda lies of the capitalists and their parasitic sycophants!

I know we can't get into all this more now, but it is very important to know the actual facts. So if people want to understand the experience of "the first wave" of the socialist revolutions, I would really encourage people to check out the research materials compiled by Raymond Lotta and others, which can be found at thisiscommunism.org and that can be linked to through the revcom.us website. I would also encourage people to study what Bob Avakian has said in analyzing the reversal of the socialist revolution in China after the death of Mao. There's a great deal of complex experience that's been analyzed and summarized and that people can get into.

But my point here, in brief, is that the capitalists who run things obviously have a vested interest in slandering revolutionaries and communists, and so they do so on a daily basis. This is not unexpected. But what I get upset about is that regular people allow themselves to be taken in by such slanders way too easily! They should really do some work and study and read up on the actual experience, instead of uncritically swallowing (and spreading) the slanderous propaganda. The materials are available, so go ahead and learn about those experiences. That's part of what doesn't go on enough, it's part of people being generally so unscientific. You hear people all the time saying, "Well, everybody knows...it was a disaster in China, or Mao executed millions of people, or the people hated socialism and it didn't work," and so on. Well, no, not everybody "knows" that! In fact some of us know that such statements are complete bullshit. Many people repeat outrageous things like that without doing a single bit of serious research on the question. That's really irresponsible in my opinion. What's true or not about this experience matters to the lives of billions of people around the world. So do your homework. Study the question. Read Bob Avakian. Dig into Raymond Lotta's analyses and compilations of research material. This material is available. So do the work and dig into the deeper substance of that whole experience. And once again here I would particularly appeal to college students: Write some papers on the Cultural Revolution in China. What were they trying to do? What were the contradictions they were grappling with? What had the old society been like before the revolution? What were the problems in the new society they were trying to resolve? What did they do that was right? Where did they make some mistakes? **Do some work.** OK?

So the primary reason there continue to be all those stubborn misconceptions about the supposedly dry, lifeless or even repressive nature of communism is because of those incessant campaigns to spread lies and distortions of the "first wave" of socialist revolutions that have been going on for the last few decades. Most people who have come across these slanders just seem to uncritically assume they must be true, and don't even bother to look into it further. Secondarily, though not unimportantly, there is also the problem posed by the fact that there actually **were** some mistakes made in the approach to running those first socialist societies and dealing with the complex problems that were involved in doing this. There were a lot of complex challenges involved in those very first attempts in history to reorganize and run a society on a completely different socialist foundation, all while having to contend with internal and external opposition from class forces aggressively wedded to the old ways of feudal and capitalist exploitation. So this was very challenging. And yes, some mistakes were made, even some serious mistakes, both in the Soviet Union and in China during those first attempts at building socialism. But there were also tremendously positive accomplishments and breakthroughs. And the mistakes can be identified and learned from in order to do even better the next time around. This is a lot of what BA's new synthesis of communism is bringing forward: Learning from both the positive accomplishments and the mistakes of the "first wave" of socialist revolutions in particular. This new synthesis, with its whole emphasis on more correctly wielding a "solid core" while allowing, and even fostering, "lots of elasticity," but still "on the basis of the solid core" does actually represent a new synthesis of communism—a significant philosophical and epistemological breakthrough and a forward leap in the whole method and approach to how to lead a revolutionary movement (today, and right up through the seizure of power), and to how to tackle the concrete and complex challenges of building up and leading a new socialist society with a thoroughly scientific method and in such a way that it really would become the kind of society that most people would want to live in.

In earlier stages of the communist movement, there was often a bit of a one-sided focus on meeting the needs of the people, meaning their most immediate material needs, for food and shelter, health care, employment—all those kinds of things—all of which, of course, are very important. But one of the things that needs to be understood more, and BA is very much a proponent of this, is that people need much more, the oppressed need much more, than just the material requirements of life narrowly defined. They **also** need science and art and culture, they need expansive atmospheres, they need room to breathe, they need room for dissent and ferment, they need room to think, they need room to do nothing. [laughs] It's a much more lively and broad understanding of what it means to correctly identify what are the objective needs of the people, even the people who are the most oppressed. Yes, they need food and health care and shelter, but they need a lot more than that. And in the past experience of the communist revolution, that hasn't always been understood well enough. It's been understood to some extent. Certainly Mao understood the importance of art, for instance. But there are some legitimate reasons to be concerned with some of the ways running and leading those first socialist societies was approached, in terms of their being too restrictive—it's the point I was making about holding the reins too tightly. For instance, I know that many artists and scientists might have been somewhat legitimately concerned that, in previous socialist societies, if they had wanted to work on certain kinds of projects, for instance in some of the more abstract realms of science and art, they might have run into opposition and not gotten support from state institutions, or enough funding and so on, especially if they were not able to explain just how their work was going to produce immediate results, results that would concretely benefit the people in an **immediate** sense. Or maybe they would be confronted by people from the bottom of society who might be narrowly arguing, "Well, you shouldn't be doing any of this, it's a waste of time and resources, how is this gonna help us live better? You can't prove to us how any of this abstract stuff is gonna help **us**." Good leadership would have to challenge such people to broaden their outlooks and recognize the societal value of allowing and even fostering a significant amount of unorthodox and unconventional "experimentation" in both the arts and in the sciences—such as the more "abstract" and rarefied endeavors often undertaken by scientists and artists trying to break new ground in their fields. Masses of people having little or no experience in these fields would have to be led properly to better enable them to recognize the value of such projects, even when there can be no guarantee at the outset that they will actually end up making a positive contribution to society in a more concrete and immediate sense. But there is no denying that providing correct and all-rounded leadership around such questions can be challenging, especially in a situation where resources are limited, where it is still a monumental struggle to meet the most basic material needs of life and where the vast, vast majority of people are still barely beginning to raise their heads above the backbreaking and suffocating exploitation and oppression they suffered in the old society, as was the case in China before the revolution.

Now, I'm not saying that, for instance, the revolutionary leadership in China, when it was actually still a socialist country, was totally unaware of this problem—of both sides of this problem. In fact they **did** promote a lot of different scientific endeavors (in advanced abstract mathematics, in medicine, and so on) that did not necessarily lead to immediate palpable benefits. They did often take a more long-term view of such things. And they also didn't crush all abstract art, or whatever. But there were undeniable weaknesses in their approaches to such things, in the direction of being too narrow and restrictive, and I think that has also secondarily contributed to some of the wrong ideas people have today, thinking that communists in power would necessarily be very restrictive and kind of crush your spirit, and that somehow the existing system of bourgeois democracy and the style favored by capitalism-imperialism ultimately provides more room for individual expression and individual rights.

But that's really not true. Under this current system there's a tremendous amount of repression, actually. Not just the most obvious repression of whole sections of society consigned to crushing poverty, exploitation and police brutality. But even repression of individual expression, and even among the relatively more privileged strata. In this capitalist-driven society there's frankly not all that much room to breathe (including little or no institutional support, and funding) for a great many artists and scientists, especially those seeking to engage in more unconventional explorations, and those whose work will not necessarily generate monetary profits for their backers and investors. You know what I'm talking about. So, the question is, **what actually is the better system, even for unleashing conscious innovation and initiative, broadly across society, and even on the part of individuals?** I would argue that socialism (ultimately leading towards communism), properly conceived of, and properly led, offers way more options for that desirable broad flowering of human initiative than the existing bone and soul-crushing system that is the current profit-driven and restrictive system of capitalism. And I really feel that if a new socialist society could be constructed on the basis of the methods and principles that define Bob Avakian's new synthesis of communism, this would really give unprecedented scope and rein to the greatest possible breadth of human exploration and experience, including by allowing for a lot of experimentation that is not narrowly tied to immediate goals and objectives, and by projecting a willingness to encourage such experimentation even when no one can tell in advance whether or not it will lead to positive results or to some kind of dead end, or whether or not it will prove to have lasting social value. Understanding that the strategic objectives of revolutionary communism—the very goal of emancipating all of humanity—actually **requires** that scientific methods be applied to unleashing and fostering such breadth and diversity and ongoing experimentation throughout the **entire** revolutionary process, and leading others to recognize the importance of such an approach—this is one of the hallmarks of Bob Avakian's new synthesis of communism.

Question: So, part of what we're getting at is the contrast between all these stereotypes that you've been talking about, that people have of communists as being gray, humorless, lifeless, dogmatic, versus the reality of who BA is and how he comes across, in the Dialogue and more generally. Did you want to speak further to that?

AS: Well, I think it comes across in many ways. And anybody who reads anything he's written or watches the films of his talks, or the Dialogue, or other things...often you'll hear people say things like, "I never realized how funny he is," or "I didn't expect this," or "he really gets the people," or "he knows exactly what my life is like." There are all these kinds of comments, and it's such a contrast with those stereotypes. For instance, if people read Bob Avakian's memoir, *From Ike to Mao, and Beyond*, they'll see it's full of stories, including some very funny stories. I mean, of course, this is a very serious guy, who has developed some very serious theory. He's very serious and deep about the analysis of all that's wrong with this society, and of how to go forward to a new world, including through the very complicated process, the very risky process, of making an actual revolution. So, yeah, he's very serious about all this, and this is not somebody who recklessly plays around with the lives of the masses, you know. He feels, I'm sure, a tremendous amount of responsibility for trying to get it right, in terms of leading people in that whole complicated process. As Mao used to say, a revolution is not a dinner party! BA doesn't tell people that revolution is going to be easy. He doesn't tell people that it won't require a lot of risk and sacrifice and that there won't be any suffering associated with it. So on the one hand, this project he's talking about, it's a very serious thing, and he takes it very seriously. And, on the other hand, he's also this very lively person who has a really good sense of humor and who can appreciate and talk about all sorts of things.

A lot of people know he's into basketball and other sports—he's very knowledgeable about this kind of stuff. He's also very knowledgeable about the law—constitutional law, civil rights and the whole legal sphere. He can tell a lot of interesting and provocative stories about that. He's a great story-teller. Anybody who's been exposed to him knows that he's a master story-teller. He's full of humor, he's very funny in a lot of ways—I think this has something to do with his overall methods and his sharp appreciation of contrast, or contradiction. That's where a lot of humor comes from, right? When people who have lots of prejudice and misconceptions imagine in their minds what a political leader, a revolutionary leader, let alone a revolutionary communist leader, is going to be like, do you think they picture in their minds the kind of person who is gonna experiment with composing and performing a rap song like "All Played Out"? Think of what it means that he wanted to do that, and that he could do that. He's not a professional rapper or anything obviously, although I happen to think that piece is actually quite good. But the point is, what does it say, what does it project to people, that here's this revolutionary communist leader, this person who says, "I'm here, and I can lead a revolution, and I can lead us into a new society on a whole different foundation, and I have all this work and evidence to back that up," and at the same time he's the kind of person who'll experiment with things like a rap song! [laughs]

He knows how to play. And he's sending a message to people: I *get* you...and he's got some life to him, you know? Many people know he loves to sing. He loves to sing, and he has a very good singing voice. I know Cornel commented on that at one point during the Dialogue.

Some people probably know that very affecting Ry Cooder song, "Borderline." BA did a cover of "Borderline" that can be heard on Outernational's album, *Todos Somos Illegales*. It's a wonderful song, it's about the border, and the immigration question, and the tormenting of immigrants from Mexico and other places in Latin America. BA sings it beautifully, with a lot of feeling, and Outernational did a very good job of mixing it in and including it on this great album. And I think it's a very good addition to that album.

What does all this say to you? What kind of a revolutionary communist leader sings songs for the people, writes rap songs and sings ballads or whatever, or writes a memoir full of funny stories about his youth and about his interactions with different sections of the masses of people, and talks, for instance, about what he learned on the basketball courts as a youth out among various sections of the people? Not exactly the dry, lifeless, humorless stereotype that many prejudiced people might have imagined, right?

One of the things that BA projects is that he has a real feel for the basic people, a real understanding of their lives, circumstances and cultures. Another way to put it is that he has science with a lot of heart. People often comment on that. People from the most oppressed sections, and in particular Black people, when they get to know him and know about him, even just a little bit, they pretty quickly sense that, Oh, this guy *gets us*. It doesn't matter whether he's white or Black, or whatever, this guy gets us, heart and soul.

So I think that's another very important thing about BA: He's obviously got a tremendously high level of theoretical development, but he's **also** got a tremendous visceral connection with the heart and soul of oppressed people, people he knows and connects with on a deep emotional level. It's not a game to him, it's not a gimmick. It's something that he really feels, and because he really feels and really knows it in his heart of hearts, often the people from the basic masses that he's speaking about, and speaking to, they quickly recognize that. **They get him getting them.**

A New Theoretical Framework for a New Stage of Communist Revolution

Question: OK. So I thought we could kind of broaden it out now from talking about the Dialogue. But just before we do, I want to echo what you were saying about how people should really go to the website revcom.us and check out the Dialogue and really take it in, and check out the film of the Dialogue once it is available. I think your phrase about how there was magic in the air is a really appropriate phrase to describe it. So I want to echo your urging people to do that. But just to kind of broaden it out, I was wondering if you could speak in a more overall way about how you see the content and significance of Bob Avakian's work, method and leadership. What is the significance of this in the world? And how this relates to the points that we've been talking about in terms of a scientific approach to understanding and changing the world through revolution.

AS: Well, I think we're talking about the most advanced revolutionary theoretician alive in the world today, the person who has taken things furthest in terms of the development of the science of revolution which started with Marx in the late 1800s and which was further developed through different periods by Lenin and Mao in particular. As time went on, and at every stage, there were some very significant new things that were learned and applied. There were some important new theoretical developments as well as practical advances. **But I really think that Bob Avakian's work in this period is actually ushering in a new stage of communism.** And that's both for *objective* reasons and *subjective* reasons in my opinion. Let me try to explain what I mean by that. First of all, there have been significant new material developments in the world even since the time of Mao, and the theoretical work that BA has done is capable of recognizing, encompassing and addressing those *objective* changes. The world doesn't stand still and we don't live in exactly the same world that Marx lived in, or that Lenin lived in, or even that Mao lived in, so the science of revolution has to remain dynamic and be able to continually develop, including in relation to these ongoing changes in the objective situation. But the reason I think that BA's work is ushering in a new stage of communism is not just because of ongoing worldwide changes in the objective situation but because of the pathbreaking breakthroughs BA has been making on what we might call the *subjective* side of the equation—in other words, his whole development of a new synthesis of communism and the radically different method and approach he is taking to the problems of advancing the revolution, both in this country and worldwide, which I feel represent a very significant advance in the development of the science itself and which stand in sharp contrast to the various kinds of wrong-headed methods and approaches which have plagued most of the so-called international communist movement for quite some time now.

BA's theoretical work has deeply analyzed, sifted through, and recast the experience of the past in a way that is actually bringing forward some new theoretical components that have never been seen before, including in relation to the concrete process of building up revolutionary movements—identifying some of the key and much more consistently scientific methods and principles that must be applied in order to do this correctly (not just here, but in other types of countries as well), the key things that have to be kept in mind all along the road to revolution, leading up to the seizure of power; and bringing forward as well, again in some important new ways, some of the methods and principles that should be applied in the approach to actually seizing power, and to going on from there to build a new socialist society in such a way that it would not only truly constitute a society that most people would want to live in, but also one that would have a better chance than past such societies of not getting diverted and turned backwards, back towards capitalism instead of forward towards communism.

But here's part of the dilemma, here's what's frustrating to me: most people today don't get any of this! They don't get the significance, literally on a world scale, of what BA's new synthesis of communism is opening up, in terms of new possibilities for humanity. People don't get this unless they actually start digging into these questions a little more seriously and actually start to grapple more scientifically with what's going on in the world, and what's actually needed.

Question: Which questions?

AS: Well, once again, the significance of what Bob Avakian has brought forward in relation to objective developments in the world and vis-à-vis some of the very wrong views and problems of method that prevail today among most so-called communists. Again, there was what has been called "the first wave" of socialist revolutions, which lasted up through the late 1970s, when capitalism was restored in China and the world was once again devoid of any genuine socialist societies. Marx really opened up that first stage of things in the late 1800s with his insightful historical materialist theoretical work on class contradictions throughout history and on the particular features of capitalist societies and the need and basis for revolutions to move beyond that towards socialism and communism, ultimately on a global scale. There was, in 1871, the experience of the Paris Commune, which was significant as a preliminary kind of attempt in which proletarian forces seized and briefly held power in Paris, but this really could not be consolidated for any length of time—there was not yet the conception, there was not yet the strategy, there was not yet really a vision, of what needed to happen to take it further. Obviously, the Russian revolution of 1917 was able to not only seize power, but to also consolidate power, and then go on to establish socialism and build the Soviet Union as a socialist state for a number of decades, before it got reversed and capitalism got restored there in the 1950s. And then the Chinese revolution, after the country-wide seizure of power there in 1949, and right up to the late 1970s, was able to take the process even further, before it too got reversed. So it's important to learn from **all** this, both from the advances and from the shortcomings.

Lenin, who led the revolution that brought the Soviet Union into being, was a very important theoretician who, among many other important theoretical breakthroughs, developed a real understanding of how capitalism had evolved into imperialism, into a world-wide system. Those were important **objective** changes in the world at the time, and some of Lenin's developments of the theory actually encompassed those changes and spoke to them in some very important ways, which I won't try to get into here. Then, by the time of the Chinese revolution, Mao advanced things yet again, bringing forward a lot of new understanding of things, like how to get started on the revolutionary road in a Third World country dominated by foreign imperialism, and what it meant to actually wage protracted people's war in that type of country over a period of time, leading up to the country-wide seizure of power. Some of Mao's greatest contributions were made after the seizure of power, over a period of years, in the course of analyzing the positive and negative experiences of the Soviet Union, and in relation to the challenges encountered while working to develop a socialist society in China. Mao's theoretical breakthroughs during those years included the analysis—the very important analysis—of what were the social and ideological remnants, the vestiges, of the old society which still exerted significant influence in the new socialist society, and his recognition therefore of the need to find appropriate ways to "continue the revolution" even in a socialist society. This was something new, that had not been previously understood or anticipated, and it marked a critical advance in the developing science of communism—a key lesson for communists to learn, and learn well, not just in China back then but everywhere around the world, and one that will be critical to have in mind in all future socialist societies. As part of all this, Mao developed critically important theoretical concepts about class relations under socialism, including the fact, that he famously popularized, that, in socialist society, "you don't know where the bourgeoisie is—it's right inside the communist party!" This is something Mao analyzed at a certain point in the development of socialist society, and he unleashed people to wage a Cultural Revolution, even under socialism, to advance things further. That was very important, and those important leaps and breakthroughs made by Mao have been deeply appreciated and analyzed by BA and have been incorporated into the new synthesis that BA has been developing ever since then. Despite all the major theoretical and practical advances and contributions of Mao and the striking accomplishments achieved in the course of developing socialism in China in the course of just a few decades, the fact that the revolution there did get reversed in the late '70s and that capitalism has been restored there was certainly a great impetus to recognizing the need to make rigorous scientific analyses of what had happened there and to develop the scientific theoretical framework of communism even further, in order to be able to handle things even better the next time around. Which is precisely what BA set out to do and the new synthesis of communism he has brought forward is very much the fruit of the work he has done in order to meet that need.

So again today, there are no socialist countries in the world. That doesn't mean there aren't revolutionaries or people talking about communism and socialism in different parts of the world, in different countries, even waging people's war in some places—or people who have done so in more recent decades. But, frankly, the international situation is a mess. The international communist movement is, by and large, a mess. And it's because of some very, very problematic lines and line differences in the international movement—some very fundamental errors that have been made in either one or another direction, and which BA has spoken to. He's helping to sort that out. But, to be blunt, he's basically not appreciated by the bulk of what has been the sort of old-school international communist movement. He's very controversial in those circles. People disagree with him a lot, because there are these very wrong tendencies and trends in different countries that get away from the revolutionary road and from the path towards genuine socialism and communism but that some individuals and organizations are very invested in holding on to, it seems. And, I mean, some people actually think he doesn't even have the right to speak about these issues because he's not from a Third World country, he's a white guy from an imperialist country. That's a pitifully narrow and pathetic way of thinking. But it's rooted not just in narrow nationalism (though that is certainly a factor), but also in the kind of devaluing of science, and of theory in general, that is so prevalent everywhere these days.

On the more positive side, I'd like to point people again to the polemics that have been written by revolutionaries in Mexico, the OCR, which can be accessed through the revcom.us site, and other things that have been written by others, polemicizing **against** some of these wrong trends in the international communist movement today and **upholding** BA's new synthesis of communism in opposition to that. Again, people should go to the online theoretical journal *Demarcations*, which can also be accessed through revcom.us. These polemics point correctly to the fact that, on the one hand, you have these dogmatic tendencies... I'll just very briefly say this: On the one hand you have these trends in the international movement that represent *dogmatic* tendencies, that argue that you only have to rigidly "stick to the fundamentals," that act as if there's basically nothing new to learn (!), despite the clear evidence that the world keeps changing in many important ways that need to be taken into account, and despite the fact that there's obviously a great need to sift through past accumulated experience in order to better learn how to avoid critical setbacks and have more successful revolutions and build more successful socialist societies. Seems kinda obvious, right? But there are more than a few mechanical dogmatic types around the world who approach revolution and communism more as a religion than a science and who therefore won't even really examine and engage these types of questions. And then there's the other kind of trend that basically says, "Well, there have been problems in the international communist movement and mistakes made in the past, so we've gotta loosen things up and just have a whole lot more elasticity and we'll be fine"—but basically they're going in circles and sort of *rediscovering bourgeois democracy*! They might as well just sign up, sign on the dotted line, to just try to obtain a few more bourgeois democratic freedoms and liberties, while essentially leaving the world as it is! This trend has very little to do with actually breaking away from the capitalist framework in any kind of fundamental sense—it often seems to be trying simply to promote the economic development of Third World countries *within* that global capitalist framework, and maybe just extract a few more freedoms and liberties, especially for the middle strata in the cities. But none of this is actually taking sufficiently into account the real core contradictions in these countries, the objective changes that have been taking place, and what it is that the broad masses of these countries actually need, in order to really break out of the overwhelmingly oppressive and exploitative framework under which they live.

Look, I realize that in this interview we can't really get into all this in detail. I more just wanted to make the point that, today, in terms of the international communist movement, well, there really is no single international communist movement. There are revolutionaries and communists in different parts of the world, and, since the loss of socialism in China, to a very large degree they've been in disarray. In fact, it was thanks to Bob Avakian that there was even a coherent analysis put forward at the time of the coup in China and the restoration of capitalism. He analyzed what actually had happened there to set things back on the capitalist path. And he helped to forge a deeper understanding of what is the correct way to unfold revolution and socialism in the modern world. But it's not like everybody decided to stand up and clap and agree with it—it's been either ignored, or very contended, and it still is, right up to this day. So frankly, it is a big problem in the world that there is not even much serious and substantial engagement and wrangling with the theoretical developments of the science of communism represented by BA's new synthesis. And it would be better if there were more unity forged on that developing foundation and basis.

Question: So, a big part of what you are saying is that the work that Avakian has done has actually carved out a new theoretical framework for a new stage of communist revolution, has actually advanced the science of revolution.

AS: That's exactly what I'm saying. And I think how much it's needed, both in this country and internationally, is pretty clear, given what is actually happening to the world and to the people of the world, and how much revolutionary change is needed. But there's so much confusion and disarray. And there are people... look, there have been attempts at developing revolutions in recent decades in Peru and Nepal, to take two salient examples. In both those cases there were some very dedicated people who made great sacrifices and fought for years to try to have revolutions in those countries, but they have gone completely off track. And the thing is, **it didn't have to end up that way**...I'm not saying that there could have been any guarantees that they would stay on track, and revolutionaries did face some very difficult conditions in both those countries. There were a lot of challenging problems that needed to be solved for those revolutions to have a chance of being successful. But the point is that there was a lot of unnecessary resistance to digging into some of the critical theoretical struggles that needed to be waged, to try to actually bring light into, shine a light onto some of the problems that were being encountered by the revolutionary struggles as the conditions in the world were changing—the conditions of the cities in the Third World, the conditions of the countryside in the Third World. For instance, the whole question of the application of solid core, with lots of elasticity based on the solid core, to those particular situations, in those types of countries, would have been extremely relevant to explore. But that kind of overarching principle is neither well understood nor even much examined or reflected on by revolutionaries in different parts of the world these days. Instead, as I was saying earlier, what you find are either tendencies towards going in the direction of brittle dogmatism and being static, stiff and controlling in a bad way; or tendencies towards throwing everything out the window by being too loose, including trying to pander to the middle strata of some of those countries and their interests—essentially advocating for what looks a whole lot like bourgeois democracy. Even if you give it the name socialism or communism, that's not what it is.

So there's a need for a whole world-wide engagement with some of these things. I really do believe, from my scientific perspective, that what Avakian has done is...he has really developed...on a number of key questions, he has really developed **some very new thinking**: about the road to revolution, about the seizure of power, about the nature of the new society that should be built up. In **all** of these dimensions he has carved out some very new thinking, identified some warning signs and problems to be avoided, and in particular he's done this by highlighting the typical philosophical and methodological errors that people tend to fall into, and by drawing out the implications of the fact that if you don't approach things with the right *methods*, there's no way you are going to be able to bring about some truly positive advances. He's shown this, and he's brought out lots of concrete evidence of this, and he's drawn on lots of historical examples to reveal patterns and show where these errors of method can lead.

In any field of science, whenever you have people who are bringing forward genuinely new thinking and really visionary analyses and syntheses, and who are critiquing old ways of thinking, old methods, old ways of approaching things, it's unfortunately often the case that, for a while at least, their work is not understood, is mocked, and reviled, or simply ignored. The history of science—all science—is full of examples of this. And it's a shame really...it constitutes a loss for humanity. In my view, every minute that goes by where Bob Avakian's new synthesis of communism is not being seriously engaged and grappled with is another minute lost in the struggle to emancipate humanity from the horrors of this capitalist-imperialist world.

What is New in the New Synthesis?

Question: Yeah, I think what you just said is a really important, a really provocative and powerful point. And I want to continue with that thread. In this interview so far, you've been talking about the new synthesis of communism that BA has brought forward, and to get a little bit more into this: What does it mean to say that there is a new synthesis of communism? Or another way to go at this is, what's new about it?

AS: Well, that's a very big question, obviously, which I can't do justice to in a limited interview like this. I would first of all point people again to the website revcom.us, where, if you go to the BA portal and you click on it, not only are some of the key works of BA in recent times featured there, but there's also a complete bibliography of core works, and you can actually access for free a whole lot of very important works by BA. He's making these very broadly available and facilitating that process. And, on that website, there are some explanations of what the new synthesis is, a brief explanation, and also some longer explanations. I think BA and the Party are making a lot of efforts to try to give this to the people, to anyone who is interested, making it available very broadly and encouraging people to check it out, making things either free or very inexpensive, trying to really make it easy for people to get into it. There are many different works, and I think it's important that people actually read BA's works. There are many books and articles and essays. There are many talks, there are films of his talks, and you can get a better sense there than what I can possibly represent here.

But I will say that some of what's new about the new synthesis of communism is, first of all, that it's much more scientific than anything that's come before. You can see this, and we talked about some of this earlier, in the ways it approaches really digging into material reality as it actually is, uncovering patterns, using scientific methods to investigate and explore ever more deeply, being willing to go to some uncomfortable places, really promoting critical thinking, being willing to look into some of the errors of the past in order to learn from them and go forward on a better basis. Look, one of the things the new synthesis has done is that it hasn't just limited itself to sorting out and distinguishing the positives and what was correct in the past experience of socialist revolutions, from the negatives and the errors that were made. It has done that, but it's done a lot more than that. It's not just some kind of cobbling together of these things. It's not just a deeper and more scientific *analysis* of the past, it's a new *synthesis*, one that is based on that deeper analysis, of how to better go forward in making revolution and building a new socialist society on a better foundation and with better methods than at any time in the past. It's actually breaking new ground in terms of sorting out and recasting the experience of the earlier wave of socialist revolution, basically from the 19th century and Marx's early development, up through the reversal of the Chinese revolution in the 1970s. Again, that's what is meant by "the first wave," and there's been a lot of deep analysis of what was correct in all these different experiences, what does or doesn't help things move forward in the direction of communism, what is actually, objectively, in the interest of the vast majority of humanity. The new synthesis has deepened our understanding of internationalism, with the concept that the whole world comes first and is the fundamental basis and stage on which all these different contradictions are playing out. It has more deeply analyzed the nature of the capitalist-imperialist system, including as it has evolved into further developed empire and has further consolidated its rule over the entire globe.

And the new synthesis has made a deeper and more correct analysis of what does it mean to meet the needs of society, to meet the needs of humanity—what I was saying earlier about going beyond strictly trying to deal with the most basic economic needs. In other words, capitalism-imperialism does exploit the working people for profit, and so on, and there is a struggle to meet the basic requirements of life; but with the new synthesis there is a greater understanding that the world we need, in order to meet the needs of humanity, has to encompass a lot more than that. It needs to meet basic economic needs, but it also has to meet the cultural needs, the scientific and artistic needs, of people broadly and in all their diversity. It obviously needs to be able to encompass and meet the needs of the most oppressed and exploited, but it needs to do even more than that. It needs to encompass very broad swaths of humanity, in all its variations and diversity. So there's been quite a bit of development in terms of a better understanding of both the nature of the problem and the nature of the necessary solutions, if you want to put it that way.

Again, a hallmark of the new synthesis is that, compared to any previous theoretical development of the science of communism, it is much more thoroughly and consistently scientific in its method and approach to everything. It puts a lot of emphasis on critical thinking and on really boldly confronting errors and shortcomings, while not denying or throwing away the actual successes and accomplishments of previous incarnations of the socialist revolution. And that's very important. It gets back to what we were talking about in terms of truth and the understanding of what truth is. What is true is what actually corresponds to material reality. That's what truth is. It's not just an idea, it's not just what you might think or what I might think. Does something correspond to the way things actually are in material reality, or does it not? What does the evidence show? You often have to be willing to dig, to explore more deeply, to uncover the evidence and get at the patterns. You generally can't just answer a question like that in two seconds. You have to be willing to look for patterns and concrete evidence that actually exist in reality. You also have to look for evidence over a period of time: You want to examine repeated examples, not just one example. You don't just want to go on very partial or limited experience, you don't just want to say, "Oh, well, this happened the other day, so obviously that's truth, or obviously that's a significant thing." Well, I don't know. Is it part of a recurrent pattern, or is it just something that occurs every now and then? I mean, what is the actual significance? You have to dig more deeply to get at the bigger lessons of life and the bigger patterns of reality. And one of the things that Avakian has done is to actually promote that kind of method. He basically tells people: Look, no matter how much you might want a better world and no matter how much you might want revolution, and you might want communism, you just can't try to twist things to fit your expectations or come out the way you'd like them to. You have to actually look for the truth of things, based on concrete evidence, even if it turns out to be an uncomfortable or inconvenient truth, and even if it ends up revealing your own errors or shortcomings. If you really want to go in the right direction, you have to be able to face up to that.

And one of the things that really distinguishes a good scientist—and I would include BA in this category—is this understanding I pointed to before, that you learn at least as much from an analysis of mistakes and shortcomings as from an analysis of successes. And again, one of the things that BA has done is dig deeply into the experience of the first wave of socialist revolution to understand where people, even the best-intentioned people, went off track, made mistakes, had the wrong conceptions or the wrong methods and approaches. And by digging into what actually happened—including some of the errors of method and approach—it becomes a lot more possible to understand what were some of the underlying causes of the restoration of capitalism, why socialism was overthrown and capitalism restored in the Soviet Union and later in China. It becomes much less mystifying or confusing. People sometimes say, "Well, if socialism was so great, how come it got overthrown, how come people didn't want to keep it?" Well, we now know there were mistakes made, and we can learn from those mistakes. But we also understand better now that one of the big problems of socialist revolution is that you make that revolution in particular countries at particular times, but meanwhile the rest of the world is still wrapped up in capitalism and imperialism; so, for a while at least, any emerging socialist country starts off embedded in an imperialist world, and this generates a lot of pressure and makes it objectively even more difficult to develop the new socialist society. That's one of the problems people have to wrangle with.

And mistakes have in fact been made in the past when trying to defend socialist societies while also contributing to expanding the world revolution, and when trying to develop the internal socialist life of a society while at the same time having to contend with all those capitalist-imperialist forces pressing in on them, antagonistically, from the outside. These are big complicated problems to have to deal with. And yes, there have been errors of method in how some of that has been dealt with in the past. For instance, there were some errors made in terms of sometimes making unconscionable alliances with repressive foreign regimes in a misguided attempt to defend new and fragile socialist societies by finessing certain international relations or exacerbating some international contradictions between competing imperialists. There were also sometimes errors of method that were made when dealing with some of the middle strata people who may have had one foot in the new society and one foot kind of back in the old society: Sometimes such forces were given too much room to exert their undermining influence, and sometimes they were given too little room to breathe and were restricted too severely.

I don't feel I can get into all this in great depth right now, but the point is that leading communist revolutions and developing socialist societies on a correct basis is a tremendous challenge, full of complexity and a great many thorny contradictions, which in an overall sense have to be handled "with just the right touch," or things can easily go off track in some very bad directions.

In my opinion the new synthesis, if it is systematically applied to such problems, provides the methods and means to unfold the revolutionary process—both before and after the seizure of power—in a much better way than at any time in the past. It really has broken new ground in terms of the process of getting to the seizure of power, in terms of developing strategy for revolution in a country like the U.S., and also in other types of countries. What are some of the key principles for getting there? What about the question of how to go about actually seizing power when conditions are ripe for this? Seizing power in an actual revolution means going up against the armed force of the state. How could you possibly do that without getting crushed? How do you do that while involving millions of people, and in a country like the U.S.? How do you do it with a realistic chance of winning? You can't just wish for it to turn out alright (!)...that's one of the big obstacles...when it comes to that stage of struggle, you're going up against very powerful forces with entrenched traditions and lots of armaments. How do you develop the work, theoretically and in terms of strategic orientation and approach, so that, when it gets to that point, people have a chance of actually winning and coming out the other end, not just having experienced a lot of loss, but with a new and better society being born and on the way to being developed?

And then there's the question of how do you nurture this new society in a way that actually moves in the direction of overcoming the "4 Alls" very concretely, in other words, going in the direction of communism. And, at the same time, do it in the way we talked about before—solid core with lots of elasticity based on the solid core. If you're too elastic, you're going to get overthrown. All these different forces of basically the old capitalist guard, as well as some newly arising capitalist-inclined forces within socialist society itself, are still going to find a lot of material basis to restore capitalist modes of production and capitalist values, if you're too loose and don't prevent that from happening. On the other hand, if you try to control everything too tightly, people broadly are going to feel like they can't breathe and things are going to chafe and grind.

Innovations are going to be stifled and people are not going to want to take a lot of initiative. There's going to be fear, there won't be enough ease of mind, and things will feel repressive even when they're not, and people just won't be very motivated to fight for this new society. So you have to get the right synthesis.

I think Avakian is breaking radical new ground on the relationship between these two aspects. There's the analogy I used earlier about riding a horse, and the two kinds of mistakes you can make: You can let the reins go too loosely and the horse will run away with you, and you'll probably get thrown off the horse that way; or, you can hold the reins in too tightly but then the horse can't even run, and nothing positive gets developed, if you follow the analogy.

So these advances, these breakthroughs in the new synthesis, are rooted very fundamentally in a rigorously scientific approach to questions of philosophy and method, applied to meeting the complex needs of humanity in the very best ways possible. Once again, in relation to the question of truth, are you going to think something is true just because that's what you're hoping it will be? Are you going to start lying to yourself and convince yourself of something that's not true, just because it might be more comfortable or convenient? Are you going to try to make reality fit your conceptions, or preconceptions, or are you going to take up scientific methods to get a more accurate picture of how reality really is? Are you going to look for immediate results in the short-term but not bother thinking about strategic objectives and how best to proceed, even now, and at any given moment, in such a way as to advance towards those overall objectives?

Theoretical Breakthroughs and Practical Application of the New Synthesis

Question: Alright, you are making the point, very importantly, that this new synthesis breaks radical new ground in terms of the process, both of making a revolution and seizing power from the capitalist-imperialists, and then continuing that revolution all the way to communism, all the way to overcoming all exploitation and oppression. And just to follow up on that, I guess one way to go at it is this: In what way does this new synthesis represent the theoretical framework that people actually need to take up and apply to making revolution and emancipating humanity?

AS: Well, look, any time you have a new theoretical framework in science, then the question is also how it gets applied. And it's very concrete. I mentioned before three key things: The method and approach to developing a revolutionary movement towards the point where you could go for the seizure of power; then the process involved in the seizure of power itself, the dismantling of the state apparatus and all the old institutions, the actual defeat of the enemy at that point, and being able to establish state power, without which you can't do anything; and then the process of starting to build a new society on a new foundation and a new basis with completely new objectives and ways of life for the people. The new synthesis has looked at the past experience of the socialist project, the first wave of the socialist project from the mid-1800s through today, through the restoration of capitalism in China as well as the Soviet Union, and has sifted through and recast a lot what can be gleaned, what can be learned from all that very rich experience, in terms of what was very correctly analyzed and implemented and in terms of what went off track, all in order to be able to do even better on all those fronts the next time around. And this is very evident in every component of the new synthesis.

First you have to get into—what is the nature of the problem? There are all these outrages and injustices that must be fought today: For instance, mass incarceration and the police murders of unarmed Black and Brown people that is so common these days in this country. This has to be fought **today**. The fact that the right to abortion has been under attack and that this right has increasingly been denied to many, many women in many parts of this country—this outrage, and all the other forms of egregious degradation and dehumanization of women as women, also has to be fought, **today**. The unspeakably brutal and vicious treatment of immigrants. The endless wars for empire, armies of occupation and crimes against humanity. The terrible and rapidly accelerating environmental degradations. **All** this needs to be fought, **today**. We should not tolerate, and we should not be complicit with, such outrageous abuses. It's all objectively part of a package, really, and it should be viewed as such: All these outrages stem from the same source, the same system. So there needs to be a lot of ongoing sharp exposure of those **systemic underpinnings** that all these outrages have in common, so that people can increasingly come to understand how deeply they are rooted and anchored in both the history of this capitalist-imperialist system and in its modern-day workings.

This is something Lenin put a lot of emphasis on—the need for communists to do a lot of exposure. He talked a lot about the need for exposure of all the outrageous abuses and injustices, to get people to understand where these came from, what was at **the root** of these problems, in the very functioning of capitalism and imperialism. So Lenin is an example of someone who had made big theoretical advances with that.

But then, there's still the question of grasping even more today, and in light of today's world, what are the implications of that kind of understanding for the strategy for revolution. We're not just talking about the new synthesis as a better way to *document* abuses and injustices. It's an actual revolutionary framework, a framework to figure out how to link up the fights that are necessary today—the exposure and the fighting against the system that needs to go on today against these abuses—how to link all that up into a bigger and more coherent process that will actually move things towards...that will create the ground and prepare the people for actually going for the seizure of power when the time and the conditions are right for this. And there are many ways in which all this is understood in a more developed way with the new synthesis. The world has changed, and the methods have been further developed in terms of this more rigorously scientific approach, which emphasizes the need for actually looking for patterns and the truth, and not just going on one or another little tidbit of experience, or partial experience in this one city, or with that one person, or with that small group of people, or whatever. Such partial and unsystematic experience is insufficient. It does not reveal the deeper features and patterns of reality. So you have to look more deeply at what's going on, and in a repeated way. And then you also have to examine the experience of trying to work on those contradictions.

For instance, on the basis of some of the theoretical advances in the new synthesis, I think that Bob Avakian's been able to develop a very rich concept of the strategy for developing a revolutionary movement and moving things in the direction of the seizure of power. There's a lot in the theory and some of the application of the concept of "hastening while awaiting" that people can read about. The concepts of how to, first of all, identify what are **the key concentrations of social contradictions** in a given society at a given time. You can't fight everything all the time. So you need a method, you need an approach, you have to apply the science to figure out what are the most **key fronts** to work on, how do you **link them up together**, what are **the forces you could unite** around that, where is the **basis for uniting** people, where is the basis for identifying that a particular set of contradictions is more important to work on today than some other set of contradictions. So, in a given period in a given society, what is the handful of concentrations of contradictions that the revolutionary forces can and should most actively work on, and work to unite broad forces to work on, throughout society, that would best contribute to concretely moving things in the direction of a revolution, that will help to build the forces, build the mental preparation of the people—you know, **"Fight the Power, and Transform the People, for Revolution,"**—preparing the ground, the terrain, the thinking of whole blocs of people, in such a way as to go in **that** direction, in the direction of an **actual** revolution.

You could take up the fight around all sorts of things. There's hardly a limit to the outrages—this system generates millions of abuses and outrages here and around the world and on a daily basis! So how do you know what to focus on? How do you know what you should link up together? How do you know what forces to go to, and seek to unite? How do you know who should be your core forces? What do you most rely on? How do you make room for and train new people? How do you lead in the best way possible? Do you just try to get people to carry out assignments, or do you put a lot of emphasis on training people in key principles and the correct scientific method and approach so that they can best contribute to the advance of the revolutionary process themselves and in turn have the tools needed to train others as well? And how do you deal with the fact that some people, spontaneously, would actually prefer to be passive and to leave the driving to somebody else? What are you going to do about that? Because that kind of orientation is sure not going to lead to a revolution! So you actually need to struggle with people to take up scientific methods themselves and to themselves help, in an ongoing way, to identify the key fronts of work and key forces to work with, the best methods for bringing it all out into the broader society, for making breakthroughs, for making concrete advances. And then, if you experience some setbacks and failures (or great successes!), how do you correctly analyze and sum up these experiences, with some depth and some substance, rather than just skipping over that and just moving on to "the next thing"? In other words, how do you sum them up in a way that you can learn some deeper lessons for the advance of revolution? If you experience some victories, how do you sum **those** up, to learn from them, so that you can actually understand more **why** something was successful in moving things forward, and what more you can do going forward? How can you develop and build off the advance? How can you use one victory as a stepping stone for more victories? Similarly, how can you learn to use even a setback or defeat as a stepping stone for doing things more correctly from now on, and possibly re-directing things, re-directing attentions, priorities, forces, and so on? There are all these very complex strategic questions that are involved in trying to concretely develop a revolutionary movement in an advanced imperialist country like the United States—to develop things in a good way towards the possibility of an actual revolution, an actual seizure of power. The new synthesis models a very scientific method and approach which can be applied concretely to doing just that, including by comparing and contrasting the correct methods and approaches to the many wrong ways people can find of going off-track!

And then, with the new synthesis, you can also apply its theoretical breakthroughs to looking in new ways at the question of how the actual seizure of power could be conceived and actually undertaken. Obviously it's not going to look like World War I warfare, with two armies lined up against each other, with set pieces or something! So while we're not at that point yet, such things need to be thought about, and while there's definitely more theoretical work that still needs to be done to further explore such questions, it is also important to study and reflect on some of the important theoretical work that has been developed in recent years in relation to the questions of the possibilities. The possibilities of actually getting in a position to have a real chance of "winning." This is a very serious subject. "On the Possibility of Revolution" is an important document to study for these purposes—on the basis of the new synthesis' overall methodological breakthroughs it is able to pose new questions in fresh ways about how could you possibly take on a force as strong as the U.S. imperialists in their homeland **in such a way** as to have a real chance of, at one and the same time, being able to unite people very, very broadly to do this, and to also having a realistic chance of "winning"—of actually ultimately succeeding in seizing state power, without being crushed.

The *Constitution for the New Socialist Republic*—A Visionary and Concrete Application of the New Synthesis

AS continues: And there are also the very important theoretical advances brought forward by the new synthesis in relation to how to start building a new society, on the correct basis and with the right methods and approaches. Here too there are a lot of ways you could go horribly off-track with all this, so it is very important to grapple, even now, with what would constitute those correct vs. incorrect approaches. There are so many things you'd have to move quickly to restructure, and some that would take more time. Of course you should have a planned economy, and you should bring into play ways of restructuring the economy so it's not geared for private profit (the way it is under capitalism) and instead it's more geared to meet the material needs of the people broadly in society. But this can't be approached narrowly or simplistically or with narrow reductionist objectives. There are many complex contradictions involved in precisely how to do that, as everybody in the past has discovered. Whom you involve, where you put your priorities, what the overall feel of life in the society will be like, and so on. The methods of the new synthesis allow you not only to recognize the core aspects of what's wrong with capitalist economies and to contrast that with the core features of a socialist planned economy that you should work to institute right away—the new synthesis also shows you how to do that in such a way as to bring along broader and broader sections of the people to willingly and consciously participate and contribute to that great societal restructuring.

Just to use one example, there's some really radical thinking in that *Constitution for the New Socialist Republic in North America* about how to constitute civil society after the seizure of power. How do you restructure not just the economic institutions and the planning, and so on, but how do you establish and apply the rule of law? And a very radical thing, relative to past experience, is that the new synthesis' breakthroughs epistemologically and philosophically have led Bob Avakian to argue that a new socialist society **should not have an official ideology**. And that the communist party should seek to lead **primarily politically and ideologically**, in other words primarily through political and ideological orientation and guidance and struggle—more in **that** way than by trying to "tightly control" every single institution of society, as seems to have been too much the case in past socialist societies. This is very important, and it's a good example of how the new synthesis has managed to internalize some of the positives of past socialist experience while also analyzing and breaking with some of the past rigidity in the approach to leading a new society. Certain critical institutions, like the armed forces for instance, would still be led by the Party, but at the same time they would be accountable to the Constitution, and it would be a violation of this Constitution, and the basic principles it embodies, for the armed forces to act against the rights of the people that are set forth in that Constitution. There would be civil institutions, and the role of the Party would remain somewhat separate from that. The new synthesis puts forward a lot of such very, very concrete and very new thinking in terms of how to approach building the new society, looking ahead to how you would structure things: the rule of law, the role of elections, contrasting elections in the current society and the future society, and speaking to the role elections should play in the overall process of the new society—these are all very concrete questions that are being deeply examined, and re-examined, on the basis of the new synthesis. How you protect people's rights, while also keeping the society moving in the general direction in which it needs to go to meet the needs of humanity, to advance toward communism. How you deal with the question of the international contributions to the revolution and how that relates to the domestic situation.

So there are many, many complex questions that this new theoretical framework actually gives you a leg up on, a good starting point, to try to very concretely deal with the challenges of building a new society, in such a way that most people would want to live in it **and** that it keeps going forward toward the goal of communism. So here I have to put in another plug for this *Constitution for the NSR*, because I don't know how much people realize what this actually represents, how radical this is! In other words, it's kinda giving us a blueprint for what to start doing "the day after." Sometimes I think, Oh, it'd be great to get to the seizure of power and actually have a socialist revolution and actually start building a new society. And then I often think, Oh boy, the day after the seizure of power—what do you **do**? That's pretty complex, running a whole society, right? [laughs] But this *Constitution*, if people look through it, just even look at the headings and topics it covers, it actually gives you such a detailed, concrete framework...it's a concrete application of the new synthesis to what the new society would look like. It really gives you a sense of where you could start, what you would start to work on transforming, and why. And I think that this *Constitution for the NSR* can be a very inspiring thing right now, today, as it can give people more of a sense of what the new society would actually look like. I think most people would actually find their place, with some genuine ease of mind, in this kind of society. I think that most people, if they really look into it, would say, "You know, I don't know about everything in here, but I think I could live in this kind of new society. I think it would deal with a whole lot of the terrible abuses of this current society overnight, and that there would be enough room for some differences and for working things out that aren't all figured out yet, and for moving things in a direction that would benefit the vast majority of people." So this *Constitution for the NSR* is a very inspiring document, which is a direct result, a direct application, of Bob Avakian's new synthesis of communism.

The New Synthesis: Consistently Going for the Truth—Rejecting the Notion of "Class Truth"

Question: Yes, I would definitely echo that point about the importance of this *Constitution for the New Socialist Republic*, and I really think people should dig into that. I was also thinking as you were talking, going back for a second to the recent Dialogue between Bob Avakian and Cornel West, that there is actually a lot that was modeled by Bob Avakian there in terms of what does it mean to apply the new synthesis of communism, both to the process of making revolution, but also as a window into what the future society based on the new synthesis would look like. There is what you were mentioning earlier, in terms of the way that BA's steadfast emphasis on going for the truth came through in the Dialogue; and the internationalism that was a big theme in the Dialogue, which is also a key component of the new synthesis. And then I was reflecting more on this point about solid core with a lot of elasticity—both in the process of making revolution, but then also continuing that revolution under socialism, transitioning to communism—the orientation of having that solid core of the science of communism, leading the whole process of making revolution and continuing that revolution, but then, as you were saying earlier, on the basis of that solid core unleashing and embracing a lot of elasticity and people and ideas going in a lot of different directions, coming from different perspectives. I felt like you really got a sense of how the Dialogue with Cornel West was an application of solid core with a lot of elasticity, and was a window into a future society where you'd be having these kinds of dialogues, you'd be having this kind of exchange, with the solid core of revolutionary communism but then embracing, and unleashing, and being enriched by people coming from a lot of different perspectives, including the perspective that Cornel West is coming from.

AS: Yes, this is important, because one of the hallmarks of Bob Avakian's overall work, the new synthesis of communism that he's brought forward, is this whole question of breaking with some very wrong conceptions that have plagued the history of the communist movement in the past, and still do internationally. Concepts like what's called "class truth." It's a very significant negative thing that Bob Avakian has completely ruptured with, completely thrown out the window. It's this very unscientific idea that, just because the proletarians are the most oppressed in society under imperialism...it's the idea that the most oppressed people in society—the oppressed minorities or the proletarians, or whatever—have some kind of special purchase on the truth. That somehow, depending on where you were born, whether you were born poor, and so on, that somehow must mean that you automatically will have a better understanding of where things should go, and what should be done. This is ridiculous, but it's a confusion that's often plagued revolutionary movements, communist movements, historically. It **is** true that the oppressed, the most oppressed in society...and that the proletarian class, **as a general class worldwide**, the class that is not in a position to own the means of production under imperialism, is clearly the class whose **objective interests** (whether or not people understand this as individuals) are most in correspondence with the direction of communism. And that's an important thing to understand—that a worldwide social class of people who objectively have "nothing to lose but their chains" are going to be the core of the revolutionary process. All you have to do is think about the difference in a country like the U.S. between some of the middle class people... even some of the progressive ones...who might sincerely want the world to be better, with fewer abuses and outrages and injustices, but who, at the same time, kinda want to keep a foot in the current order, in order to be able to keep benefitting from some of the benefits and "perks" that this system can still often afford **them**, in their own day-to-day lives...compare that to the people at the very bottom of society for whom daily life is usually a horror, and who, objectively, don't have anything really worth preserving out of the current set-up. So, spontaneously, those people at the bottom of society may be more ready to move in a radical revolutionary direction towards the new society. But that doesn't mean they automatically have a better handle on the truth or understand things better, just because of their position in society! And one of the things Bob Avakian is always stressing is that we have to be open and willing to learn from all corners and all spheres of life, and all different kinds of people, coming from different social positions and having many different perspectives on things. This approach was clear in the kind of engagement that took place during the Dialogue, and it's a hallmark of his overall method and approach.

Let me give you another example, a very **negative** example that quite a few people may be familiar with, the Lysenko example... from back when the Soviet Union was a socialist country. It's one of the sorrier pages of the history of the communist movement. I cannot imagine this happening with Bob Avakian's new synthesis of communism. You know, if you're familiar with the story, basically in the Soviet Union in the days of Stalin, they had a very big need...they were confronting some complex challenges and were trying to meet a very big need...there was an urgent need to produce more food and to rapidly increase agricultural production in order to feed people, because there were famines, and so on.

And there was this person Lysenko, this scientist—well, he was an agriculturalist—and he was apparently very much in favor of socialism and communism. So you might say he was "with it," in terms of personally wanting the socialist society to continue and advance. But scientifically...look the guy was a terrible scientist because he had a completely wrong understanding about biological evolution, and he actually held some pre-Darwinian beliefs in the supposed inheritance of acquired characteristics—the long since disproved wrong view that if an individual plant or animal acquired certain characteristics in the course of its lifetime, those characteristics could somehow be passed on to its descendants. This wrong notion had by then been scientifically discredited for decades, but Lysenko still clung to such false notions. But because Lysenko was politically in favor of socialism, his wrong scientific views were given a hearing. And you actually had some scientists at that time in the Soviet Union who had a much better and more correct handle on scientific facts and principles. But many of them were more bourgeois, or petit bourgeois, in terms of their class origins or recent ways of life. And some of them were not so much in favor of this new radical regime, this new radical system called socialism. Maybe they kind of liked the old ways better, at least for themselves, or maybe they had a mixed view of things under the new society. In any case, they tended to be more critical about the new system and about the leadership of that system. But they had **better science**! So they said, no, this is not the way to boost agricultural production—what Lysenko is saying is wrong because this is not how biological evolution actually works, and you're not gonna increase agricultural production by applying wrong scientific principles! Well, here's where we get to one of those "truths that make you cringe" in the early history of the socialist project: Even though Lysenko was arguing for this complete junk science, leaders of the new society listened to him and allowed him to implement some wrong-headed and disastrous agricultural policies, just because he was someone who wanted to promote socialism and communism. Instead of relying on sound scientific principles and methods that were by then well-established, Stalin and others in the Communist Party leadership, like Lysenko himself, fell into practicing **instrumentalism**, trying to "fit the truth to a desired outcome," rather than proceeding from the actual truth, from reality as it actually is, and working on its actual unevenness and contradictoriness in order to change things in the desired direction. The leadership rejected what was being argued by these other scientists, at least in part just because some of them weren't very enthusiastic about socialism. Well, maybe they weren't so much for the revolution, but they were **right** in terms of their scientific understanding, and Stalin and the other leaders should have listened to **them**. Lysenko was in favor of the revolution, but he was completely wrong in his scientific understanding, in a very damaging, destructive way. And the fact that the leadership messed up in evaluating that—because they held on to a wrong and unscientific method and approach to assessing the actual truth of something (regardless of where it comes from)—this caused serious setbacks in agriculture and more generally set back science in the Soviet Union in an overall sense. And to this day this Lysenko story, about a grievous error in method on the part of the communists and its very negative real world implications, turns some people against communism altogether, because obviously nobody would want to live in a society where that kind of thing might happen on a regular basis.

But look, the better synthesis would be: Let's learn from such mistakes. Let's learn the lessons deeply and well. The leaders of the new society in the Soviet Union were trying to figure out how to feed the people in a time of hunger and famines. That was their intention, and this was not an easy problem to resolve, even with good methods. But it did make things much worse that they had a very bad philosophy and method on this question of science. Bob Avakian's new synthesis of communism would never go for that. Do you understand what I mean? I don't know if I'm explaining it really well, but I think it's a really sharp example, because the new synthesis recognizes that you don't have a special purchase on the truth because you were born poor, or because you were born as part of an oppressed minority, or because you were born in a Third World country, or because you were born an oppressed female, or because you are in favor of socialism and communism. None of this gives you a special purchase on the truth. The truth is the truth. It's what corresponds to objective reality. And **anybody** can work on discovering the truth, no matter where you were born or what background you come from.

The question then becomes, are you really going for the truth? Are you basing yourself on actual, scientifically determined, evidence-based truth when you set out to develop plans and policies? If so, we can and should learn from you, no matter who you are. Bob Avakian's always stressing this, that we can learn important truths from all sorts of people. Even sometimes from people who are in the enemy camp, who are enforcers of the system and defenders of the system. Even they can sometimes come up with insights or new knowledge that we can learn from. You just have to check out what the evidence is...critically examine the proof and the evidence and the actual patterns of the reality. And if something is true, then that's good, you can use it, you can use it as part of the revolutionary process. You certainly don't want to base things on a faulty understanding of things, just because you are hoping it will provide you some kind of shortcut or fit your preconceptions of the way things are supposed to be.

Bob Avakian: A Very Rare Combination—Highly Developed Theory, and Deep Feelings and Connections with Those Who Most Desperately Need This Revolution

Question: Before turning to the next question, I wanted to emphasize, once again, that these points you were talking about in the new synthesis—the development of internationalism, the strategy for revolution and seizing power, and what is concentrated in the *Constitution for the New Socialist Republic in North America* about the kind of new society envisioned in the new synthesis, and the whole scientific method and approach of that new synthesis—all that was reflected and modeled by Bob Avakian in the Dialogue with Cornel West. And I think we should get even further into the significance of the new synthesis and what this breakthrough represents, and the need for people to take that up. But, before that, I just wanted to back up for a second to something you mentioned a little bit earlier, which I thought was important, which I would describe this way: that there is a really rare combination in BA's work and leadership, of doing highly developed work in the realm of theory while, at the same time, as you were saying earlier, speaking very directly and viscerally to those most brutally oppressed by this system, those who most desperately need this revolution. That's very rare, and I wondered if you wanted to say a little bit more about that.

AS: Well, yes, mainly to very much agree with you that this is extremely rare, and it's an extremely precious combination. And this gets back to the point I was making earlier. I think **a lot of people don't have any conception, really, of what revolutionary leadership is**, especially at a high strategic level like this. I spoke earlier about the misconceptions that people have, that sometimes when they imagine what a revolutionary leader is, they think about an agitator on the street, or somebody leading a demonstration, somebody down on the ground, at that kind of level. And that is an important component of the revolutionary process. But that's not the same thing as the overall strategic leadership that is taking on the whole history up to this point, and all the many complex interconnected contradictions, setting key points of orientation and guidance, and breaking with mistakes from the past, while identifying the correct trajectories from the past that need to be further fleshed out, understanding how to prioritize and strategize, in order to advance the whole complex process towards a whole strategic worldview and vision on a global scale. **That is a whole other order of things, which requires very complex theory.**

And you do have some people historically—you have some intellectuals who grapple with points of theory, but who are often not grounded very well in the actual conditions of the people, or the needs of the people in different sections of society, especially the ones that they're maybe less familiar with, the people at the bottom, consigned to the bottom of society. And then you have some people who are maybe very familiar and comfortable at the basic levels of society, are very familiar and very comfortable with the basic masses of the oppressed, and who can speak in very compelling ways to their conditions of life, but who, because of their own conditions sometimes, have been deprived of the ability to do a lot of synthetic analyses and theoretical development. And none of these are fixed categories, you know. Everybody has to strive to develop on **both** fronts—**the theoretical and the visceral**.

But what you've got in somebody like BA—and I can't think of another example in the world where this is as developed—is you do have this incredibly developed theoretician who **also** has this incredibly acute visceral sense of the conditions of life of different sections of the people, especially of the most oppressed at the bottom of society. This is not some kind of abstract question for him, this is clearly something he feels very deeply. And the people themselves also recognize that he feels this very deeply. That's why I made that point earlier, that **"people get that he gets them."** And I think that, for a revolutionary leader, that's a very important characteristic. But you can have people whose heart is with the people, but who don't have the theoretical, strategic commander kind of development and abilities. Or, again, you might have some people who have some theoretical abilities, but who are kind of...

not intentionally, but they're just not that familiar with the people, and the most oppressed in particular.

So to have **that combination**—somebody who is really connected to the heart and soul of the deepest level of the most oppressed of society, and at the same time is able to handle a very wide range of subjects on a high theoretical level, who has actually advanced theory on multiple fronts and brought in some new theory to guide the practice of today, and of the future—I don't know what else to say, except that it's an incredibly rare combination that people should consciously reflect on and appreciate.

Again, some people, if they conceive of what a revolutionary leader is, if they try to imagine that, they think of the leader of a demonstration or something. They don't understand the role of a strategic commander or of a theoretical developer of new theory, and so on. The other mistake some people make is that they think, maybe, of a revolutionary leader as somebody who works in a room—what used to be called an armchair Marxist—somebody who's sitting around concocting theories kind of divorced from reality, or maybe even pretty good theories based on analyses of past history or something, but who doesn't know what's going on too much in the society and the world today, and doesn't know how to guide the practice. You have some people who can make some theoretical arguments, but who have no idea how to guide today's revolutionary movement in a coherent, systematic way that builds towards an actual revolution.

And that's the other kind of mistake—the idea that BA is an armchair theoretician who works in a room surrounded by books, and that's the end of it. You know, **he is actually providing ongoing practical leadership to the *entire ensemble* of revolutionary practice in this period**, the whole development of the movement for revolution. The kinds of questions I mentioned before—like what should be the main fronts of struggle today; what are the key concentrations of social contradictions that exist under the system that the other side, the system, cannot resolve and that the people need to work on, to push on, in order to make some breakthroughs and unite people in a revolutionary direction and make some advances concretely that way. There are many decisions to be made: what to take up; what not to take up; how much weight to give to different aspects of the practice. I think people, sometimes, they have...I think **if you're not scientific, you don't understand the connection between theory and practice very well**. The RCP takes on many different fronts of practice—for instance, the struggle against police murder and brutality and stop-and-frisk and mass incarceration. The RCP has played a leading role in developing some of that. Well, where do we think that comes from? Where does the initiation of that, where does the guidance to that come from...not to every nitty-gritty detail of daily practice, but to the core conceptions of why and how this should be taken up, how much emphasis to give to it, what are some key principles and methods to have in mind? It comes first and foremost from BA. Or questions involving the oppression of women, to take another example. It's not like somehow BA is developing some new thinking about the future socialist society but is somehow not involved today in very concretely providing guidance to the initiative against the patriarchal degradation and dehumanization of women—everything from pornography and rape culture to the denial of abortion rights, the denial of reproductive rights, and so on. Of course he is involved in providing leadership to that! These are not issues that somehow only other people are figuring out how to expose and organize resistance around without ongoing input and leadership from BA himself.

When it comes to leading the revolution, BA is both a very developed and visionary theoretician and a very sharp and experienced down-on-the-ground practical leader. He is very much **both** things—that's the point I am trying to get across. And yes, that combination really is very rare, and it is also very important, and very precious.

Again, this is something I think people often don't understand—that if you're the leader of a revolutionary party and you are a key leading theoretician who is bringing into being a new theoretical framework for the revolutionary process and for the building of a new society, that doesn't mean you're just kind of "out in the clouds" somewhere with some abstract thinking, even some good abstract thinking, but without any connection to the development in practice of the day-to-day struggles. Quite the contrary. My understanding as a scientist of the process involved, and what is evident in BA's writings, which I do study carefully, is that he is very, very much integrally involved in every aspect of the development of the revolutionary movement of today on all these key fronts. These are not fronts that are being led just by other people. There are of course other leading people who are taking on very important and critical responsibilities vis-à-vis these fronts, but they are not doing this in isolation and separately from the leadership, the overall strategic leadership, that is being provided by BA and by the central leadership of the RCP to **all** of it, **to the entire ensemble** of work in this period.

The Charge of "Cult"—Ignorant, Ridiculous, and Above All Unconscionable

AS continues: There is a lot that is not well understood, especially if you don't have a thoroughly systematic scientific approach to reality in general, about **the relationship between leadership and led in a revolutionary party and a revolutionary movement**. I'm sorry, but I always have to laugh—even though I know it's a serious slander and can't be taken lightly—but whenever I hear of anybody accusing BA and the RCP of being some kind of a "cult," or something, I have to laugh, because that's the most ridiculous thing! There is "strategic commander leadership" being provided by Bob Avakian to the entire process of developing the revolutionary movement today, to developing the methods and approach to be able to get closer to the point where an actual revolution would be possible, to starting to work out the features of the beginning of the new socialist society—all the things we've talked about before, in terms of the breakthroughs of this new framework for revolution, which is very concretely being applied. And in a revolutionary communist party, there's a **collective** process, and this itself is a complex process that does require other people to play significant roles, at different levels. There are some people who play very important leadership roles in their own right. My experience with leading people in the RCP is that there's a whole bunch of very different, very strong personalities, who are the furthest thing from slavish followers of a cult figure! [laughs] That's just not what's going on. There is line and direction that is being hammered out, in an ongoing way, including through a collective process. And the reality is that it's pretty clear that BA is miles ahead of everyone else, in both theoretical development and in the application of the science to practice, to the development concretely of the revolutionary movement. But then there are also some other people who play critical roles, who significantly contribute to the overall process, and take initiative, and who are part of the process of analysis and summation and synthesis. There's a back and forth between leadership and led, and in a healthy environment this happens at many different levels, right down to the newer people coming into a party, and to people who are outside a party but who can still make some critical contributions, which have to be recognized and encouraged and brought forward and not stifled, and which in turn can feed into the process of further developing the basis for advancing the overall strategy for revolution.

So, I don't know if I'm expressing myself very clearly here, but it's a whole strategic—as opposed to just tactical—approach to a revolution. It's not just approached in piecemeal fashion. It's not just do a little bit here and do a little bit there, fight this outrage and fight that outrage, and maybe take on this and criticize a little of this and a little of that, or whatever. It has to be **all pulled together in a single direction**, even though it's made up of a lot of different complex interacting parts. There has to be an overall approach...there has to be leadership that can provide a broad overview and direction, an overall methodological direction and orientation and guidance for this whole period of history, and which can also provide concrete direction to how to best carry out all the different components involved in building a movement for revolution in the right ways. And I strongly feel that the new synthesis of communism developed by BA is able to do precisely that—both parts of that.

But applying the new synthesis to tackling the problems of the revolution today doesn't mean that people should be trying to **passively** implement guidance and leadership that they just hope will be handed to them on a platter by higher-up and more experienced leaders. In fact, it is a big problem whenever supposed revolutionaries adopt that kind of passive attitude! **It is very good to be disciplined, but it is not good at all to be passive.**

Passivity is certainly not something that BA would ever encourage among followers of his new synthesis, as should be crystal clear to anyone who actually looks into his works. In fact the whole method and approach of BA's new synthesis of communism insists that people **at every level** of the communist party, as well as the people, at every level, who are involved in the broader movement for revolution, should all take up very active roles and go out into society, like teams of scientists, actively engaging and interacting with reality as it actually is, keeping in mind and proceeding back from the longer term strategic objectives, while finding many different and creative ways (in line with those strategic objectives) of concretely advancing the movement for revolution today—accumulating forces and organizing people to fight the power, and working on transforming the thinking of blocs of people, all while bringing out to people broadly: Why there is a great need for an actual revolution to really get beyond the outrages generated by the current system; why there is in fact a material basis and genuine realistic possibilities for revolution in a country like the U.S.; how it is that a much improved society could be built up on a completely different socialist foundation, once the current system of capitalism is cleared away. The idea is to get more and more people, from all walks of life, to seriously engage all of that. And, on the basis of all that, creating new conditions and changes in the terrain, changes in the people, changes in the objective conditions, that in turn can provide some new bases for further advances, including theoretically. There is no question in my mind that such a complex process requires leadership, including from the highest, most experienced and developed levels; but it **also** requires, and must be continually fueled and enriched by, a great deal of creativity and conscious initiative on everyone's part, at every level of the party and of the broader movement for revolution. And BA himself is continually stressing the importance of this, of **both** aspects of this. Things must be led, and led well; and on that basis a great deal of creativity and initiative must be unleashed. This is yet another expression of his key principle of "solid core, with lots of elasticity on the basis of that solid core."

So there is this continual back and forth that goes on, between different teams or different levels of interpenetrating leadership and led. And again, it would be completely ridiculous to call this a "cult"—it is actually very much the opposite of a cult. Of course it will always be the case in life that some individuals might **prefer** to be passive—we've probably all known some people like that. But in the context of a party or movement for revolution, such people really should be struggled with not to be passive. Because you're never going to make a revolution worth making unless everybody pitches in, in different ways but with the right spirit and in a unified and disciplined manner. The revcom.us website has been promoting this call to "get organized for an actual revolution." You know, you do have to organize people. You can't just put out a few bright ideas here and there and expect to make a revolution. [laughs] In order to be tightly disciplined and organized, leadership should in fact be followed and respected—but not passively, not slavishly. People should be critical thinkers...if something doesn't sit right with them, they should ask questions; if they have different views of things, they should be raising them, they should be saying what they think is not right and why, and so on. But there has to be a disciplined approach to carrying out work in a revolutionary direction, in order to advance things, and to provide a richer basis for things to be summed up and for further guidance, further direction to be given from a strategic level.

Leadership: Does It Stifle or Unleash Initiative?

Question: I think part of what you're pointing to also is this question. There's a widespread conception in society that leadership stifles initiative. But does it stifle initiative, or does it actually unleash people and unleash initiative?

AS: There's no question in my mind that in any sphere, including in the natural sciences, but also in this scientific communism, good leadership is always seeking to unleash initiative, but this has to be done in a disciplined, organized manner. Think about it. If you were doing a project in the natural sciences, and you were trying to get people to tackle a particular problem or particular set of questions, and you were trying to get people to work together collectively on the project, it wouldn't work out very well if everybody just took off willy-nilly in any old direction of their own personal choosing and started off by applying completely different sets of working assumptions and theoretical frameworks and templates to the problem right at the outset, in a kind of anarchic manner. Certainly the best collective natural science projects I've ever been involved in **have been led**, and have unleashed creativity and individual initiative **on that basis**. Have definitely unleashed individual initiative and creativity and all sorts of individual contributions, but **on the basis** of initial, and also ongoing, good scientific leadership. I've learned a lot from that kind of leadership/led interaction, when it's correctly conceived of and unfolded.

Leaders of scientific teams in the natural sciences are generally not shy about providing leadership! [laughs] Such leadership is often provided in the form of such things as: identifying key problems to resolve and the questions to focus on at any given time; delineating core guiding principles and methods, based on prior accumulated knowledge and the most developed experience in a given field or sub-field of natural science; articulating sets of working assumptions and hypotheses and the basic theoretical framework to take out into the world, with which to poke and probe and seek to transform reality. In short, one way or another, **good science projects are led**. And I think most natural scientists understand on some level that, no matter how many people might be involved in a project or how much money or other resources you might have at your disposal, you're never going to get anywhere or make any real progress in advancing scientific understanding or in resolving complex scientific questions or problems if you proceed to work on things in an anarchic manner, absent a sound scientific base and ordered structure **from which** to proceed, including to best enable you to encounter and explore completely new things or concepts that were previously completely unknown or not yet understood. Without this underpinning in an ordered base and structure, in what is essentially the best possible grounding in the most developed scientific theory available at the time, you won't be able to even pose the right questions or correctly and systematically analyze and synthesize what you encounter as the project unfolds. And you certainly therefore won't have a very good basis for either further contributing to the accumulation of new scientific understanding or to transforming material reality in certain desired directions (by curing a disease, figuring out how to protect an ecosystem, or whatever) if this also happens to be your objective. Right? Well, all this is definitely the case in the natural sciences, but the very same principles also apply if you're trying to apply science to understanding and transforming a society, including by applying scientific methods to the complex process of making a social revolution. This process, too, needs to be **led**, and individual creativity and initiative and all sorts of individual contributions definitely can and must be unleashed, but this can best be done **on the basis** of sound scientific leadership. This ongoing and very positive interplay of leadership and led is especially important to keep in mind and actively contribute to if you're actually trying to **change** things in the world, for the benefit of the many, and not just trying to indulge your "self" or just your own individual interests and proclivities.

So again, good leadership should constantly try to consciously unleash initiative, and you can't make a social revolution without unleashing a tremendous amount of conscious participation and conscious initiative on the part of growing and increasingly diverse numbers of people and types of people. But the problem is, it's a two-way street! You have to take some responsibility for this process yourself. You know, when you look into Bob Avakian's materials, you see that he's constantly calling on people—inviting people and struggling with people—to enter into the process, to actively engage things and not remain passive. But some people resist that, even some well-intentioned people sometimes resist that. If somebody says, "I don't wanna get a headache trying to wrestle with complicated questions, so just tell me what you want me to do, and I'll do it"—that's no good. You have to struggle with them. You can't make a revolution that way! You have to say, "No, that's not right!" You have to do some work yourself. You have to think about what's right. You have to study the orientation, the direction that's being provided by leadership. You should try to evaluate it critically while, yes, at the same time, going out with it, taking it out into society, into the material world. In other words, you should work to take things out into society, systematically and in a disciplined way, **on the foundation of** the leadership guidance that's being given, but then you should also be sure to contribute to systematically analyzing and reporting back on what you're encountering, what you've been doing and what you've been learning, so that all this can feed into and enrich the overall collective process. That's a scientific approach to unfolding revolutionary practice.

Question: I think this also gets at the question of the role of outstanding individual leaders, and in particular the role of BA, because, just to go back to what you were saying a minute ago, you were making the point that BA is miles ahead of everyone else, both in terms of the development of the theory and also in terms of the application of the theory, the practical application of the theory. So I wondered if you could talk a little bit more about the kind of the relationship there is, or should be, when you have someone like BA who actually has that advanced understanding and how that actually contributes to unleashing that initiative, like you're saying, in an organized and disciplined way.

AS: Well, look, to me it's a question of basic scientific materialism to understand that everybody's not going to have the same abilities, everybody's not going to have the same level of understanding, just to state the obvious. And I do think that anybody who has some honesty, some principle and integrity, who actually does a close study of the work, of the extensive body of work, that BA has developed over a number of decades, is going to be struck by the fact that...look, whether you agree with it or not, if you are an honest person with principle and integrity you should be able to pretty quickly recognize that this work is of a whole order of magnitude beyond what prevails in the society generally, beyond what has passed for so-called leadership in the so-called political movements, or even revolutionary movements, in recent history. BA's one of those rare individuals that comes around once in a great while, as the world changes, as society changes...in the context of these objective changes and developments, sometimes individuals emerge who have particularly developed skills and abilities and some very new ways of thinking and some trailblazing approaches to leading and transforming things in some very new directions. This is true in every sphere. It's certainly true in the natural sciences, and in things like sports, music or all the many other spheres of art and culture. Just think about it for a minute and I'm sure you can come up with quite a few examples from those different arenas. For a number of different reasons, factors and influences that come together in sometimes unexpected ways, there are simply individuals who periodically emerge with particular qualities and particular skills and abilities at a given time, and who kind of rise above everybody else in their field. And the crime, frankly, is when other people in society are not willing to recognize that, or even seriously check it out and take a good look to see if that's in fact the case. It took a long time before a visionary pathbreaker like John Coltrane could be recognized and appreciated in the field of jazz for instance, just to take one example. At first, people covered their ears and complained his music was just too dissonant, and uncomfortable to listen to, and oh yeah, his solos were too long! [laughs] Seriously though, especially when somebody is dedicating their whole life to trying to make a better world for all of humanity, you would think that this would require that people at least give them a good solid hearing—and actually read and study what they've brought forward—and not just engage in facile dismissals, without even really investigating what it is that has been brought forward. And it galls me to no end that most of the people today who engage in "facile dismissals without serious engagement" of BA and his body of work, are people who themselves have nothing of substance to offer in terms of any kind of serious programs and solutions to the world's complex problems. We should keep asking such people: "What's **your** program? What's **your** strategy? What's **your** solution to the problem of the recurrent horrors generated by this system?" And, well, if you don't have much to say about any of that, if you don't have much serious substance to offer in terms of strategic plans and programs for systemic change, then maybe you should have the decency to shut up for a while and get to doing some serious work yourself to at least more thoroughly explore and engage the substance of the work done over a number of decades by someone who in fact is proposing a substantial, multi-layered, radically different and yet coherent and scientifically grounded in reality, vision and plan for the future.

You don't have to agree, but it's unconscionable in my opinion not to seriously delve into this work. Unless, of course, you just don't care. Which I guess is a big part of the problem in the current self-absorbed society: too many people care more about cultivating personal views and opinions that they can personally feel good about, than about exploring methods and approaches, and strategies and programs, that might actually enable millions... billions...of people to free themselves from horrific conditions of exploitation and oppression that weigh down their entire lives. That's what we're talking about freeing people from. What are YOU talking about?

And I think that anybody who does look into BA's work seriously, and who's basically honest, is going to end up saying, "Oh, OK, well, I didn't quite realize the complexity of what was involved, and all the thorny contradictions that are being wrestled with, and all the ruptures with some past incorrect methods and approaches that BA has been leading, and I wasn't really familiar with the ways in which he's been arguing for a whole new framework, on many different fronts, in terms of the process of how to build a revolutionary movement, what kind of revolution to make, how to have a chance of winning, how to develop a new society...I hadn't realized he'd been working on all that, with this much depth and substance...." There's so much that's new, and so much that's rich and complex, that any honest person willing to set aside prejudice and misconceptions and really explore his work with an open mind will likely quickly recognize this and may well be intrigued and provoked to explore things further.

But then the question comes up, what about everybody else? OK, there's BA and everything that makes him stand out as unique, but what is everybody else doing? Well, for one thing, "everybody else" isn't the same either. There are different levels of revolutionary communist leaders, people with different strengths and shortcomings, with different abilities, making various contributions to the revolution; there are also different levels and abilities of participants in the broader movement for revolution; and there are of course brand new people, people who come from a wide variety of different backgrounds, entering into all this for the first time. But what I want to stress here is that **everybody** has a role to play in the revolution, **everybody has something they can contribute to the process**. That's something that BA always promotes and encourages. It is important to understand that this is a revolutionary movement that is **not just** for intellectuals, for those who may have the training to handle complex abstract theoretical writings; and that this revolutionary movement is also **not just** for the basic people who are the most exploited and oppressed at the bottom of society (though it most certainly **is**, especially, for them). This is a revolutionary movement that is truly for **anybody** who feels that the whole world, including this U.S. society, is currently overflowing with absolutely unacceptable horrors and injustices and outrages, and who wants to put a **stop** to those things, and work for a better and more just world. A world where you can actually **advance toward the emancipation of all of humanity**—not just work for the emancipation of your own group, or just your own "identity," so that your own people or your own identity can get a chance to lord it over other people—instead, work concretely for the genuine emancipation of **all** of humanity.

And there is a place for anyone who thinks and feels like this to participate in this movement, and there is a place for everyone to learn and develop more as they go along. I think it's very important for people to realize that they are actively being **invited in** to be part of this process, including directly by BA himself, and to know also that the revolution, the revolutionary process itself, ultimately cannot go very far forward without them. That's a simple fact.

So, in my opinion, people should do more conscious thinking about **not only the responsibilities of leadership *but also* the responsibilities of the led**. The responsibilities of the led vis-à-vis leadership, as well as the responsibilities of leadership vis-à-vis the led. I don't think enough people, even in the revolutionary movement, give enough conscious thought to that. It's not just a question of getting people to "do a bunch of stuff" and to just participate in various actions, or various initiatives, as important as all that is. Again, the approach can't be one of just trying to get people to "do a lot of stuff." It's getting people, on every level, every kind of person who wants to be part of this and who can be part of this, to bring their ideas, their experiences, their questions, their initiatives, and to help further identify and better develop the ways that they and others like them can be actively part of all this—participating in finding their place in the revolutionary process, contributing to it, and working on developing themselves, as well as others, in order to help raise everyone's level in an ongoing way.

Here's something else to think about: What kind of a revolutionary movement would it be if individuals came into it at a sort of beginning, elementary level at one point, but then years go by and they seemed kind of stuck in place, like they hadn't developed significantly more theoretical understanding, more practical skills, more scientific methods or the ability to take on significantly more leadership responsibilities? This would be a real cause for concern, and something that would really need to be addressed by leadership in order to transform this situation, right?

On the other hand, you know, you look at an example of somebody like Wayne Webb (Clyde Young) who came from the basic masses of Black people and who did quite a bit of time in prison in his early years. People can learn about his life and contributions at revcom.us. My point here is that he learned, he studied while he was in prison. He became a revolutionary. He got into BA, followed the leadership of BA, and himself became a high-level leader in the RCP, the Revolutionary Communist Party. That's the kind of inspiring transformation that actually can and does happen, especially under this kind of leadership. You know, many prisoners actually do a lot of serious study, many seriously study the revolutionary process, and they're a tremendously precious resource. When I was writing the *Evolution* book, there was a great deal of important feedback that came from some of the people who are incarcerated, and who I guess were pretty motivated...they had the time to read and study, but it wasn't just that—they also had the motivation...because I guess they understood that this kind of learning wasn't just about learning a few, even a few very important, scientific facts and principles—some of them seemed to really "get" that all those questions of scientific methods and principles, the ones that are repeatedly hammered at in that book, have a great deal to do not only with understanding material reality as it actually is (in all its dynamic contradictoriness and unevenness) but also a great deal to do with understanding that, **within those very contradictions** (whether you are talking about a biologically evolving system, or a societal system that will change through conscious human interventions) lies **the very basis for that reality to change**, or to **be changed**. So some people, including some people who live under the difficult conditions of incarceration, do seem to get why all this matters, profoundly.

Again, whether people come from very difficult circumstances in life, live in very oppressed conditions in the inner cities, or are even incarcerated, and whether they have had the privilege of fancy education or have received very little education, there's room for everybody, and there's a need for everybody to get involved. Anybody who basically says, "Enough! I'm not going to tolerate any more of this—these police murders, these rapes of women, these endless wars, this destruction of the planet, these hounding of people across borders—I'm not going to accept a world like this any more, I'm not going to agree that this is the only way, or the best way, the world can be," anybody who sincerely feels this way, and who is serious and honest about being willing to learn, to study, to discuss, to actually participate in the revolutionary struggle, will find a place in the movement for revolution, and should themselves, right from the start, be learning what it means to lead.

And if you do want to learn to lead in a revolutionary movement, here's a tip: Study how Bob Avakian leads. Study what he models in his books and writings, in his talks, in the films, in things like the Dialogue he did with Cornel West. Study what he does: how he talks to different sections of the people; what he focuses on; how he brings out the problems in society; how he brings out the solutions; how he doesn't pander or cater to people's backwardness or misconceptions, but, rather models that Malcolm point: He tells people **what they need to hear**. Even if they don't necessarily wanna hear it, he tells them the truth, and what they need to hear. Study how Bob Avakian struggles, repeatedly and right down on the ground, with his audiences, with various kinds of audiences, to bring them to a better place, to a higher level of understanding. And then go ahead and work on doing this yourself, in your revolutionary work, in your discussions with family and friends. Learn from those methods, and be part of the revolutionary process in that way. And generally, keep thinking about that important relationship between leadership and led. This is something for everyone to reflect on who wants to be part of the revolutionary movement.

Why Is It So Important, and What Does It Mean, to Get Into BA?

Question: OK, picking up on your point that there's a role and a need for everybody who finds the state of the world intolerable, to get with this movement for revolution—with that point in mind, and picking up on the point that you just made, I wanted to ask you to speak directly to people with different levels of familiarity with BA, whether they're just finding out about him or whether they've been familiar with his work for a while. And the question I wanted you to speak to is, basically what it means on different levels for people to get into BA and get with his leadership? Why is this so important?

AS: Look, in a lot of ways, it's pretty straightforward. Whatever kind of background you come from, whatever your position in society today, if you're the kind of person who feels that there are a lot of things that are really messed up about the way the world is, the way society is; if you're completely outraged and unwilling to tolerate for a minute longer a lot of the more outrageous abuses and injustices of the society—and there's no shortage of these that different people become aware of—then follow your conscience, first of all. Follow your conscience and follow your convictions, and follow the trail to see what does BA say, given that we're saying, look, this guy's been doing decades of work, analyzing why these problems keep happening, what they are rooted in, what's the fundamental reason these terrible things keep happening in this society, what could be done about it.

So, "get into BA" means, first of all, don't stand aloof when you see these great injustices and this great suffering of the people from various directions. Get involved. Get involved in fighting and exposing these abuses, in joining with others, get organized to fight these abuses, to expose these outrages, to make it clear you're not tolerating them. And, as you become part of this movement, this movement for revolution, this movement to fight the power, to transform your own thinking and the thinking of your friends and family, and others, on some of the key questions of the day, as you get more involved in that, at the same time as you're doing that, go deeper. Go deeper. Be like a good scientist. Get into BA, because by getting into BA you are going to learn a lot about the deeper source of the problems and what the actual solutions are, or are not, and what can be done about these problems.

A good place to start, if you're new to things, besides something like the Dialogue, I would highly recommend that people watch, and watch with their friends and families, and so on, the film *Revolution—Nothing Less!* It's six hours of BA talking, analyzing the deeper sources of the problem in society and the solution. There's also an earlier film, *Revolution: Why It's Necessary, Why It's Possible, What It's All About*, that is also full of very helpful material which gets very clearly into why a revolution **is** necessary and why it is **possible**, even in a country like the U.S., and what kind of revolution should we have, what kind of society should we bring into being. And you can get a copy of *BAsics* and just start reading it anywhere you want, read some of the quotes and the short essays that are in there that give a feel for some of the range and the depth of analysis on some of these questions, and why this system cannot just be fixed with a few tweaks, **why it can't just be reformed**.

One of the most important things that it means to get into BA is to get into the deeper analysis and exposition of precisely why you cannot reform this system, and why capitalist imperialism as a system needs to be completely dismantled through a revolution, an actual revolution, and replaced with a new society built on a completely different economic and political foundation and, correspondingly, very different social values and ways people can relate to each other and function in society.

So, getting into BA—well, there's not just one way to get into BA. There are many different ways. If you're new to things, I would recommend the Dialogue, *BAsics*, *Revolution—Nothing Less!* I would recommend going regularly, at least once a week, to the revcom.us website and exploring not only *Revolution* newspaper on a regular basis, but also going to the other portals on the website: the portal that has BA's works, and talks about the new synthesis; the portal that talks about the Party he leads, why there's a need for a revolutionary party and what that involves and why people should be talking about joining this revolutionary party; and the portal that talks about what's going on in the movement for revolution and these different initiatives to fight mass incarceration and police murders and brutality, and to combat the restrictions on abortion and combat the degradation of women through pornography and rape culture and in other ways, and the fights around the environmental degradation and against these imperialist wars, and so on. There are a lot of practical things, practical initiatives, that people can get into, and people can get involved in one, or more than one, of these initiatives. But getting into BA means, at whatever level and wherever you start, trying to really get into what is he saying about why you can't fix things just one thing at a time, and why you can't fix things by trying to reform this system. And it means getting into what is the relationship between fighting particular abuses and outrages today, and being able to get to the point where you're in a position to have an actual full-out revolution to dismantle the existing system and reconfigure the society on a completely different foundation.

There is something that has been on the revcom.us website recently which gives a short definition. It's called "What IS An Actual Revolution?" I'm just going to read it here:

> An actual revolution is a lot more than a protest. An actual revolution requires that millions of people get involved, in an organized way, in a determined fight to dismantle this state apparatus and system and replace it with a completely different state apparatus and system, a whole different way of organizing society, with completely different objectives and ways of life for the people.

"With completely different objectives and ways of life for the people." I just want to stay on that for a second. That's a very important thing, how a society is organized. What does the system that governs the society—what is it aiming to do? Under capitalism-imperialism, it seeks to meet the needs of the capitalist-imperialists, in terms of being able to develop and sustain their empire, increase their profits, compete successfully with other capitalist-imperialists. It has very little to do with meeting the needs of the people, either the basic requirements of life or these more intangible things like art and culture and science. It is geared to meeting the needs of the capitalists, the imperialists. The whole state apparatus, the police and the armed forces, and so on, these are state institutions that are geared to buttressing, supporting and strengthening this system and its objectives, its goals.

And a real revolution is a process which actually involves getting to the point where you can fight—where millions of people can be involved in fighting—to dismantle, to break down and completely break apart the existing institutions of the system, and replace them with completely new institutions and new organs of power, and new ways of setting up the economy, and all the things that flow from that, including all sorts of ways that the people relate to each other, and all the ideas that flow from that. It's a radical transformation in everything, from the way people live, to the way people relate to each other, to the way people dream and aspire to things—all of this undergoes a real sea change when you have an actual revolution.

And if you get into BA, this is what you should be looking for—you should be looking for his method and approach, his analysis of **why** things are the way they are. Why do police keep murdering unarmed Black and Brown people in this society? Why is this continuing what has been in existence since the days of slavery and Jim Crow? The "lynching culture" is now the "police murder of people culture"—it is an extension of the same thing. BA does a very deep and insightful analysis of the whole history of this and of how this is so woven into the fabric of this system that you can't just wish it away, and you can't even protest it away. You should protest, you should have strong fights against this, organize with other people, like the movement that's been developing since Ferguson, a movement which started around Trayvon Martin and then built up around Mike Brown and Eric Garner, with people coming forward all over the country in different cities, and in cities around the world, to denounce this kind of thing. It is extremely important not only to continue this, but to actually expand it and have it grow and be stronger. It is part of organizing the people and strengthening the people.

And the same can be said around a number of other key concentrations of social contradictions, these key outrages in society, these **contradictions that this system cannot fix**. They cannot fix the oppression of Black people, of Brown people. They cannot fix the way immigrants get blocked from borders or turned away or tormented or denounced as "illegals," as if any human being were ever an "illegal" human being. It cannot fix the way women are degraded and treated as less than full human beings in this country and all around the world. It cannot fix the environment in any sustained fundamental way—not as long as we live under a system that is driven by a constant search, and a fierce competition, **for profits**, not as long as those are the rules of the system running things. The rules of the game for capitalism-imperialism are that capitalists are constantly competing with other capitalist-imperialists around the world to divide the world, and for plunder and pillage, and to increase their profits; and if you don't play that game, and beat out others, you go under. So even if an individual capitalist wanted to have a more enlightened position, they can't really do anything about it. The system is set up to meet the needs of that system, not the needs of the people.

So when you get into BA, you should be reading and listening and checking out all the different things that intrigue you and interest you, and talking to other people about it. You don't have to agree with everything, just check it out, do some work. Get together with people, read things, listen to things, watch films, discuss things, and develop your own understanding, at the same time that you're out there fighting and denouncing and exposing the injustices of everyday life as they're going on right now, which BA is definitely encouraging people to do. At the same time that you're doing that, keep going deeper, so that you really start to understand why these terrible outrageous abuses are **built into the fabric of this system**, and you can push back for a while, but ultimately you cannot totally get rid of these abuses until you get rid of the system itself and replace it with a completely new system. BA makes very, very deep and insightful analyses of all of that: why you **need** a revolution; what is the **possibility** for a revolution; what is the **basis**—on what basis, even in a powerful imperialist country like the United States, with all its military and stuff, is it actually possible; and, if you succeed in seizing power, then **how do you build a society that was worth fighting**

for, that you'd want to live in, as opposed to yet another bad system. **All** of that is in BA's works.

In that statement I was reading about "What IS an Actual Revolution?" it gets into **the link between fighting the power today and building for an actual revolution**. It says, "Fighting the power today has to help build and develop and organize the fight for the whole thing." "The fight for the whole thing"—if you don't get into BA, you're going to be missing a lot about how to link all these different things together and understand their root causes, which are deeply built into the fabric of the system, and you won't know how to fight for the whole thing. And you have to fight for the whole thing in order to have an actual revolution. And this statement ends, **"Otherwise we'll be protesting the same abuses generations from now!"** I don't know about you, but I don't want that, three generations hence...people have made that point about Emmett Till, that people protested what happened to Emmett Till, the lynching of Emmett Till way back in the 1950s, and here we are protesting what happened to Eric Garner, to Mike Brown, to Oscar Grant, to Ramarley Graham, to Amadou Diallo, to Trayvon Martin, to Tamir Rice...the list goes on and on and on... Right? How long are we going to be doing the same thing?

So, yes, first of all, we should definitely be protesting, but we **also** need to go deeper and be more scientific and more organized and more unified, and we need to get smarter, frankly, about how we take on these things. Bob Avakian has developed **a whole strategy for the whole revolutionary process**—not just for one corner of it, but for all the different components of it—in this country, in a country like the U.S., as well as having some very important insights for the development of revolution in other countries, including other types of countries, like Third World countries that are under the domination of imperialism, where it is necessary to work in somewhat different ways for an actual revolution.

So all that is some of what it means to get into BA. You know, just do the work. Go to the revcom.us website. Get *BAsics*. Get *Revolution—Nothing Less!* Watch the Dialogue. Read BA's memoir, *From Ike to Mao and Beyond*. Listen to some of the cultural things. Listen to "All Played Out." Play the "Borderline" song on the Outernational album. Watch and listen to these things **with other people**. There are many, many different angles and many different ways to get into this. Look, BA has spent a lifetime, he's spent decades developing all of this work. You're not going to be able to catch up, you're not going to be able to "get it all," in a few weeks or a few months or even a few years. But make a start, and then follow it up. Ask questions if things aren't clear enough. Struggle with other people about their misconceptions. Go to your friends and family and talk to them about what you're learning. And, by the way, expect to be mocked, ridiculed, criticized, to be told you don't know what you're talking about! Expect push back, OK? But **don't give up** when that happens. Do the best you can in answering things, on the basis of what you've been learning, but when you run into things that you don't understand well enough, go back to BA, go back to his materials, dig in a little deeper, talk to other people who know more about this, ask them for help, so that you can keep spreading this among the people.

The idea is that, if you had hundreds more, thousands more, tens of thousands more, discussing, debating what BA has been bringing forward, what he's arguing for, his analysis, what he says the nature of the problem is, and what the nature of the solution is, then it's not that everybody would immediately agree, but we'd all be so much better off—if everybody were having that kind of discussion, instead of just sort of turning their backs on the problems of society and the problems of the people and just cultivating their own little "self," or maybe actually trying to fight some of these abuses but in a way that is sort of like being stuck in a narrow little cubicle, where you're just taking on one issue, or one corner of one issue, but you're not seeing **the bigger patterns**, and you're not linking them up to the **other** egregious outrages in society, and you're not understanding that there is actually **a strategy** to get out of this mess once and for all. There **is** a strategy. This is why you should get into BA, and this is how to make a start.

A Hopeful Vision—On a Scientific Basis

Question: What you're saying about what it means to get into BA does cut right to the heart of what is the strategic orientation of

"Fight the Power, and Transform the People, for Revolution," and the role of theory within that. And I think a lot of people, when they first kind of awaken politically and come forward, they think that the process of changing the world, going to protests or going to demonstrations or taking action in that sense...all of which, as you're saying, is extremely important, an extremely important part of the revolutionary process, but then I think part of what you're pointing to, when you're talking about going deeper, is that this, by itself, is not enough, that people need theory and they need **this** theory in particular.

AS: Right. Sometimes people look around at all the messed up things that people do to each other, even among the people, even among the most oppressed—the way people kill each other, or the way people degrade each other, or things like that—sometimes people look at this kind of stuff and they say, "This is hopeless, how could we possibly get together to make a revolution and build a new society? People are too messed up for that, we're too messed up for that." Nonsense. That's nonsense, because the way people think and the way people behave is deeply influenced by the prevailing culture and by the prevailing forms of organization of the system, what they encounter day to day and from the time they're born, how they're being influenced and shaped by that.

But I'll tell you, as a scientist, as someone trained in biology—we don't have time to get into all the scientific details now, but I can assure you, **there is no such thing as fixed and unchanging human nature**. There is no unchangeable human nature. There just is no such thing! If you look throughout history, human beings have always had tremendous potential to take up different worldviews, different behaviors, different attitudes, positive or negative, in how they relate to each other. Even what is defined as right or wrong, what is considered "socially acceptable," changes a lot depending on what period of history you're in, what kind of society you're in, what are the prevailing conditions and the prevailing traditions in a particular system. None of that is fixed and unchangeable. It can, and does, change. So, really, what we should be doing is talking more about what would actually be the vision of a truly liberating society, of a society on a global scale that would really be emancipating for all of humanity. What would that look like? What would some of the key features of that be? And BA speaks to that, too, very much so.

And if you keep that in mind, and you talk about this vision of what the future could be like and the potential people have for **changing themselves in the course of fighting the power**...see, this is something people don't understand enough, that when people lift their heads and band together, unite together with other people to fight the power, to fight the oppressors and to learn in the course of doing that more deeply where the problems come from and what the solutions might be, they change themselves in the process. We all go through that. We all go through this process of not just learning about things, but of changing our views, our misconceptions, our bad behaviors. This is something that has happened again and again, especially in periods of revolutionary upsurge.

Question: So it seems like another really important dimension that you're highlighting, in terms of why people need to get into BA and what that means, is that people actually need to understand that the world doesn't have to be the way it is, and people don't have to relate to each other the way they do.

AS: Right. One of the things about BA's whole method and approach, his whole scientific analysis of the source of the problems and the nature of the solution, is that it's actually **a tremendously hopeful vision**, a very hopeful analysis, but one that is **not based on hype**. There are people who'll try to sell you a bill of goods and tell you, "Oh, you can have pie in the sky, you can have a better life," on the basis that they're trying to con you or sell you something, or whatever. You see that all the time. But this isn't what BA does. He doesn't sugarcoat it. He tells you, Look, it's hard, OK? We've got very powerful oppressors whose system is set up to maintain the oppression of all sorts of different sections of the people, and exploit and oppress all over the world. That's the way the game is rigged in favor of their system. And then you have the masses of people, everything from the people at the bottom of society to the somewhat more privileged middle strata—people who are not really part of the ruling class of the system, but who've kind of bought into it somewhat, even while many of them don't like the way things are—and many people as individuals are pretty messed up, and they've got a lot of wrong understandings. People don't understand how things work, why things got to be the way they are, and they don't understand their own behaviors, their own ways of relating to people or why people...a lot of times, people think that the people who are messing them up are their family members or their friends or their colleagues or their neighbors, or something like that. **They don't see the hand of the system shaping people's wrong thinking and bad behaviors.**

And, if you become part of the revolutionary movement and you take up BA's method and approach to things, one of the things that will happen is that you will develop the tools to help transform your own thinking and behaviors, and the thinking and behavior of other people that you relate to. Because you will be understanding and putting forward a vision of the way the world could be, of the way society could be organized, in a much more inspiring and hopeful way, in a way that would create a world that you would want to live in. It wouldn't be perfect, but it would be so much better than the world that we have today!

And the more you become convinced, on a deep basis, on the basis of science...you know, it's not like a preacher telling you a few things to make you feel better, or something...the more you see the scientific evidence of what **the basis** is for change to take place, and how we can actually influence things in that direction, how we can drive change in the direction of an actual revolution to build a society that would be worth living in...the more you get into that, you become a different person yourself, and you are able to help other people raise their sights and become different people—become more the "people of the future" than of the present, if you want to look at it that way. People who can become role models for others. And, again, people from the bottom of society have a particularly important role to play in this, a particularly inspiring role to play, in kind of leading the charge in that respect. But other people also very much need to change, and will change as well—students, youth from the middle strata, and so on.

And how people relate to each other in daily life—that can change radically, too. Gang fights, are you kidding me? C'mon, we gotta do better than that, we don't want these gang divisions where people are killing each other. These are all **our people**. We need to be strong together and uniting to fight a common enemy, not fight each other. Also, men trashing women, degrading women—c'mon people, let's stop doing this! Let's focus on the real problems and the real enemies, and let's get together and organize and unite to fight those things.

We've seen it before, we've seen it in previous periods of revolutionary upsurge. The people get better. The people get smarter. The people get more lofty. They dream bigger, and they act in accordance with these bigger dreams. It's a beautiful sight. And that's a lot of what BA is actually giving us the tools to accomplish.

Question: And when people are getting into BA, they should be studying his method. Could you talk a little bit more about that?

AS: Well, that's something people are not used to doing. Not just uneducated people, but even very educated people, in this kind of society, they're often taught in schools just to look for conclusions, you know, look for "factoids," for little bits and pieces of information, or for recipes or directives about how to do this or that. But people are not taught to look, consciously, at **methods**. They are not taught to examine and to compare and contrast different methods and approaches to understanding reality and changing reality. And the way we started this interview, we talked about what is science, and above all what's important about science is the **method** of science. It's a tool. It's not just a set of answers. It's a toolkit. A scientific method is a toolkit that you can take and that you can apply to every question that comes up in material reality, whether in the natural world or in the social world. It's a toolkit that can help you understand why things are the way they are, how they got to be this way, whom do they benefit, why is it continuing like that, how could they be changed, what are the obstacles to changing them. These are questions that get posed with a scientific method. So when you listen to BA or you read his books and articles, don't just wait for the punch lines. Don't just look for the conclusions. Actually look for the development of the arguments, look for the method, which is a very scientific method, a method that is based on evidence. Again, I have to stress this: **Science is an evidence-based process.** So look for the discussions of the repeated patterns of evidence that make the case, that provide the proof, for why this system is rotten at its core and cannot be salvaged; that make the case for why a revolution—an actual revolution to dismantle this system and replace it with a completely different system on a different economic and political foundation—is absolutely essential if you want to deal with these core problems, these horrible fundamental problems that cause so many people so much unnecessary suffering.

So again, look for the method. When BA talks about how complex a problem it is to figure out how you can unite all these different kinds of people...we're talking about the reality that you can't have a revolution with just a few dozen people here or there... you have to involve millions and millions of people, and they're going to have very different views, different questions, different interests, different beliefs, and so on. They're not all going to be the same, they're going to be incredibly different in some significant ways. So **on what basis** do you bring them all together and point them in the same direction and guide them in the same direction? Look for the method. Or how can you deal, in the right ways, with the contradictions that will exist in the new socialist society, when you have the old capitalists that are trying to come back, and some new capitalist forces that emerge right within socialist society, and they're trying to overthrow socialism and bring back capitalism, how are you going to deal with all that? Look for the method. And how should you handle the fact that among the people, once again, you've got all these differences and these different viewpoints, and people are pulling in different directions, and they want this and they want that...this is a very complicated process to handle correctly. Again, look for the method. You need science to sort it out, to know what to prioritize, what to emphasize, what to struggle with the people about, and how to struggle with them.

For instance, look at religion among the people. BA will tell you, religion is **an obstacle** in people's thinking, it actually gets in the way of people taking up a fully scientific approach to reality, to understanding it and to transforming it. So, should we be shy about struggling with people about that, just because so many people care about their religions a lot? Or should we tell them the truth, and struggle with people to give that up, and understand where this came from, and how religions were invented by people, and **why** they were invented, and get into whether religion is more an obstacle or more a help, in the struggle for revolutionary change? Let's get into things like that.

Similarly, let's talk to men who say they want to be revolutionaries, but who don't really want to change the way they relate to women. The men who still want to keep that one little corner of privilege for themselves, who still want to act as oppressors and dominators in relation to women, who are hoping to hold on to that misogynistic male supremacy corner of their lives. Don't do this! What kind of revolutionary are you gonna be if you try to hold on to a little piece of acting like a dominator and oppressor yourself? Let's talk about that. Let's struggle about that.

There are a lot of complex contradictions. You can't make a list of every single contradiction, between the people and the enemy, or among the people, that needs attention, that needs struggle...but you can identify **key concentrations of these contradictions**, you can prioritize some of the big ones, and work on those especially.

So, what do you do to involve the people as thoroughly as possible in the revolutionary process? You give them scientific tools so that people increasingly start to act like scientists out in the society, so they learn themselves to identify patterns, and to ask questions like: Where did the problems come from? What are they rooted in? What is causing problems today? Why can they not seem to change even when people protest? Why do the problems keep coming back, over and over again? What can be done about it? Give people scientific tools to dig into these questions, and which they can use to better understand reality and also to transform reality.

And BA is—you know, it's an irony that a lot of natural scientists haven't really recognized this yet—but BA is the most scientific person around today when it comes to dealing with social matters, with the organization of society and the problems of society. You know, I often feel like calling out not only to the natural scientists, but to the popularizers of science—people like Peter Coyote and Alan Alda, or someone like Seth MacFarlane, who helped to produce the new *Cosmos*, the follow-up to Carl Sagan's wonderful initial series, people who are not scientists themselves but who put a lot of effort into popularizing science—or people like Ann Druyan and Neil deGrasse Tyson, people who are very committed in different ways to popularizing and promoting scientific methods and scientific thinking among the people, but who seem to have a blind spot when it comes to...well, maybe not all of them, but I'm saying some of them have a blind spot when it comes to questions of social change, and of revolution and communism. They often have many misconceptions, in large part because they haven't applied their own scientific methods to really looking into all this, to really looking into what it's about and what's actually been developed theoretically, and to really give it a good critical evaluation. People too often have a sort of knee-jerk response, and a lot of times they just uncritically buy into some old assumptions and basically the crude propaganda promoted by the people running society today, so they don't even critically examine the work that's being done. I feel like saying to many of the natural scientists and popularizers of science, "Look, with BA, you have somebody here who's really good at applying consistent evidence-based scientific methods to both the analysis of current society and the pathways for change of future society. How can you not want to seriously examine and check this out?" BA is using the methods of science to better bring to light such things as: how human society is organized; what are the prevailing characteristics and features of modern-day capitalism-imperialism; why the basic laws of functioning of this system cause it to continually generate and regenerate certain types of problems that are deeply rooted in its material underpinnings, problems that are genuine horrors for the people, causing a great deal of unnecessary suffering for millions and billions of people here and around the world; why the prevailing system itself, because of its material foundations and ways it must objectively function to maintain and extend itself, is objectively, materially, incapable of fundamentally ever resolving these contradictions; and how, still on the basis of evidence-based science, we can actually analyze not only what's wrong with current society but also identify the material basis for transforming society in a much better direction, and ultimately in ways that would benefit all of humanity. Why would any scientist with a conscience not be interested in exploring the application of sound scientific methods to doing all of that?

We can talk about that some more, but something I'm acutely aware of, especially because of my scientific training, is that **the**

basis for change, of any given thing, resides within the very contradictions that characterize it. That's true in the natural world. It's true for biological evolution, for instance—the material basis for ongoing evolution can be found in the genetic variation within a population of plants or animals. That's a form of contradiction. I see the natural world in terms of contradictions—contradictions that are part of its motion and development, and that continually bring forth the raw material, the material basis, for change. And it's like that too in human society. The raw material, the basis for transformation of human society, resides in the underlying contradictions found within that system, and with good scientific methods and tools you can understand that, analyze it, and on that basis understand where you can and should focus your efforts, what contradictions you can and should particularly work on, push on, in order to move things in a revolutionary direction, and achieve a more fundamental transformation of society.

Serious Engagement with the New Synthesis—The Difference It Could Make

Question: The statement "On the Strategy for Revolution" from the Revolutionary Communist Party, which people can and definitely should read at the website revcom.us—that strategy statement talks about the orientation of right now working to bring forward, orient and train thousands of people in a revolutionary way, with those thousands reaching and influencing millions of people even before a revolutionary situation, and then in a revolutionary situation, leading those millions of people. It seems like part of what you've been saying is that getting into BA is a really essential element of that process of bringing forward, orienting and training thousands of people in a revolutionary way who are reaching and influencing millions now and then in a position to lead those millions in a different situation, a revolutionary situation.

AS: Yes, absolutely. You can't have a revolution without revolutionary theory, without scientific methods that can clarify what is correct versus incorrect revolutionary theory, and what is the direction things should go in, or should not go in. You don't have a chance of having a successful revolution if you don't take up these kinds of methods. And BA models these kinds of methods all the time. All people need to do is pay attention, listen and read and discuss.

Question: I want to return to what we were talking about a little while ago, in terms of what it means that Bob Avakian has brought forward a new synthesis of communism, and I wanted to ask you about the basic difference it would make for, not just a few people, but waves of people, to take up this new synthesis of communism that BA has brought forward, the difference that would make in the world right now.

AS: Well, look, the new synthesis of communism, it's either correct or not to call it a new synthesis, and people are welcome to grapple with this and to dig into it, and should do that in order to recognize that it actually **is** a new synthesis. But what that means is that there's a whole leap, a qualitative leap, in our human understanding of the stage of history that we're in, and what is the material basis underlying the contradictions, the problems that exist in the world today, including in this society, but everywhere else as well, and what can be done about it. There has been what people refer to as "the end of a stage of communism," the end of the first wave of the socialist revolutions, and **there's an opening now for the beginning of a new stage**. But unfortunately today, there are very few genuine communist revolutionaries anywhere in the world, at least at this point. Very few. So it's not like the new synthesis of communism brought out by Bob Avakian is the fashionable method or the fashionable ideology that zillions of people around the world are taking up at this point. Quite the contrary. People are fighting **against** it. People are trying to turn away from it, mainly by ignoring it, neglecting it, refusing to dig into it with any kind of depth and substance, which we'll get into some more in a minute, I'm sure.

But it's an outrage that people refuse to engage this. When somebody has done this much work and spent their whole life developing this framework, you should first of all do some deep study of the framework instead of just superficially dismissing it. If you dig into the new synthesis deeply and then still end up having some serious differences with it that you can actually flesh out and argue about with substance—well, OK, then, let's go ahead, let's have those discussions, let's have that kind of serious engagement. But most people, even some people who consider themselves revolutionaries or communists, they'll just kind of lightly dismiss it, without any serious engagement and without even knowing what it's about. They haven't done the work. They haven't read the articles and the books. They haven't listened to the talks and watched the movies. But they still feel entitled to just reject and dismiss it, telling anyone who will listen that it's no good. To me that is unconscionable. Because what is right and what is wrong in these matters will ultimately have bearing on the lives of millions and billions of people. So **it matters**, and deserves to be seriously engaged and evaluated.

Or, in the very few instances where people have actually tried to do a critique of the new synthesis, they have typically grossly distorted it, revealing that they do not really understand it, particularly in terms of method, and the application of that method to a number of critical questions. Once again, people should go to the online theoretical journal *Demarcations*, which can be accessed through the website revcom.us, where you can find substantial polemics that thoroughly examine and refute these critiques of the new synthesis and discuss what the new synthesis actually is, and why it is so crucial.

Then there are people who say they are anti-communist, simply because they're full of prejudices and misconceptions. Because somebody at some point in their lives told them, "Oh everybody knows communism failed, people hate socialism, capitalism is the best possible system." You "just know this" why? You just know this because somebody somewhere told you that "everybody knows that!" and you bought into this, on blind faith! The only reason you supposedly "know" that communism is no good is because the authorities, the ruling authorities, have been promoting that kind of anti-communist propaganda for a very long time. But face it: You don't even really know what you're talking about. You'll just repeat the slanders about China and Mao, for instance, but have you seriously looked into it to see if it's really true? Do you even know what kind of contradictions, what kinds of big social problems, they were trying to resolve? What kinds of social needs they were trying to meet? Do you even know what great accomplishments they actually managed to bring about, to the benefit of the vast majority of people, and in just a few short decades? Do you know what their starting point was, the problems they were dealing with when they seized power, and do you know what the old society was like in China, under feudalism and colonial domination, what the old ways of doing things were that were so brutal and oppressive and that they were working so hard to dismantle and get away from? Be honest: You don't really know any of that, do you? Not really, or in any case not with enough depth. Because if you did, you probably wouldn't make such facile dismissals of the struggles that they waged. If you actually studied all this deeply and with an open mind, you would know more about, and you would be talking more about, the phenomenal transformations that they were able to accomplish in that society under socialism, in the days of Mao's leadership. Once again, you can start learning about this by going to the thisiscommunism.org website, which you can link to through the revcom.us website, and checking out some of the substantial work that Raymond Lotta has done on this, and reading some of the analyses that Bob Avakian has done about the restoration of capitalism in post-Mao China, the nature of Chinese society before the revolution, what Mao and the revolutionary party in China were trying to accomplish, and did accomplish, that was extremely positive, and what difficult contradictions they were grappling with, including within the party.

You know, a lot of these people who just give you the one or two sentence dismissal about how "everybody knows that this was a disaster, or everybody knows Mao killed millions of people, or everybody knows...." **You don't know shit, OK**, to be perfectly blunt about it. You don't know shit, because you haven't done the work. And especially those of you out there who are supposedly intellectuals, academics, knowledgeable people, educated people, and scientific people—give me a break! Because you wouldn't be able to hold forth for ten minutes about what they actually did accomplish, about what they were trying to accomplish, about what problems they were running into, about how they tried to resolve them, about the complexity of the contradictions they were wrestling with, about the many instances in which they handled things the right way, the instances where they made mistakes...you can't really talk about any of this with any depth or substance because you haven't really bothered to look into this seriously, so you basically don't know shit. And yet you somehow feel completely free to repeat and spread a bunch of canned propaganda slanders that supposedly "everybody knows..." and that you have just swallowed whole, uncritically and on blind faith. Shame on you! This makes me mad. Don't tell me that it's acceptable to just dismiss a whole huge human social experiment, one that concretely accomplished many wonderful things, and that secondarily had some shortcomings and problems. Don't tell me it's OK to just arrogantly wave your hand and dismiss it without having even really looked into it seriously and with an open mind. That's just so socially irresponsible. That is so unconscionable. That is, frankly, disgusting.

And again with very, very few exceptions, in the international movements, or in this country, people won't even discuss and dig into BA's new synthesis. They'll just make snarky comments and remarks like, "Oh BA, everybody knows...that's some kind of a cult. Everybody just blindly follows him around, or they think he's the greatest thing since sliced bread, or whatever." Once again, this is totally socially irresponsible. Somebody has spent decades of their life working on the problems of society, working on the problems of revolution, working on the problems of how to actually fight in a way that you could win, but also fight for a goal that is worth winning, and in a way that is consistent with that goal, so you don't end up with something worse or just as bad as where you started off. Someone who has grappled with great depth and richness with all these issues, and has actually brought forward actual evidence, people—concrete evidence, OK?—repeated patterns of evidence. So you know, bring me the proof. Bring me the evidence. If you claim this stuff isn't right, bring me the proof of why it's not right. Do the work. Bring the evidence. Because BA **does** bring the evidence. He **does** approach things scientifically. He doesn't say things because he just "feels" like saying them and because this comes from just some kind of personal "belief." He doesn't go on belief, he doesn't base himself on any kind of "belief"—instead he bases what he says and argues for on concrete and demonstrable scientific **evidence**. So how 'bout you try to do the same thing for a change?

On the positive side, it would make a big difference if more and more people became actively committed to learning about the new synthesis—to honestly, with principle and integrity, dig into some of its key principles and key methods and approaches, and, on that basis, critically evaluate it. And then promoted even more widespread discussion about all this in society more broadly. You don't have to agree with everything, you don't even have to know everything about the new synthesis to recognize that broader society-wide wrangling with the new synthesis could have great social value. Again, if somebody has spent 40 years developing a new synthesis, I obviously wouldn't expect somebody just encountering it to start "representing" for the new synthesis, or to be able to do a full criticism of it, or whatever, after just a few months. But even someone who has only recently been exposed to this work in a beginning way should be able to recognize that there's something very substantial and very serious here that at least deserves serious engagement, and societal debate. Whatever people end up thinking about it, people should be engaging and discussing it seriously, studying it. Undergraduate college students should be writing papers, graduate students should be doing Ph.D. theses, on this new synthesis of communism. Regardless of where you ultimately fall out on it, BA's body of work is substantial, it is deep and profound, and it demands to be taken seriously, and to be analyzed seriously and to be debated and discussed very widely throughout society.

So even if you had just that going on, it could make a big difference in a positive sense. My understanding is that a big objective of the BA Everywhere campaign, for instance, is to broadly promote precisely that kind of broad societal engagement and debate about this new synthesis. The goal is obviously **not** to try to get a bunch of people to be blind followers of BA—"blind followers" could not contribute to advancing this project, this conception of revolution. It requires **conscious engagement**. OK? So the idea is not to get a bunch of blind followers but to promote widespread engagement with the new synthesis broadly throughout society, to get as many people as possible to become familiar with BA's new synthesis of communism, to discuss it and debate it, study it and wrangle with it, and bring in their own insights, and bat things around with others who are doing the same thing, all as part of a process. And those people who become convinced that this new synthesis really is on track, really is where it's at, in terms of analyzing the source of the problems of society and analyzing what the solution should be—those people who become more and more convinced of all that should become **active fighters for the new synthesis**, spreading it to every corner of this society, as well as around the world, taking it out to all sorts of people, to the people in the neighborhoods of the inner cities, into the prisons, into the halls of academia, into the scientific institutions, to people in the arts and other cultural spheres...in short, everywhere!

The only people who...sometimes it's frustrating because sometimes you get the sense that the rulers, some of the ruling class figures, are the ones who pay the most attention to these theoretical breakthroughs in revolutionary strategy and vision and conception and method. Of course that's because they're trying to defend their system against this! On the other hand, sometimes the people who need this the most, who suffer the most, or who have the most criticisms of this system and the way things are, who may complain on a daily basis about this problem or that problem, sometimes they can't be bothered to actually do the work to learn about this whole vision and conception that's been worked out over 40 years and more, and that is actually claiming to have a way out. Now, if you end up deciding that no, you don't like it, this is not the way out, fine, then make your case for that, and go do your own thing. But meanwhile, don't say that you're part of struggling against the problems in society and all these different outrages, but at the same time make a point of refusing to seriously check out what BA's been arguing for. That would be like refusing to take advice from a doctor who has decades of experience and expertise dealing with a medical problem—why wouldn't you want to hear what that doctor's got to say?!

Question: Yeah, I definitely want, in a moment, to come back to this point about the snark and slander, and the haters, and the dismissal without engagement. But first I wanted to go a little bit more at this point about how the new synthesis is a theoretical framework to initiate a whole new stage of communist revolution, but at the same time it's facing a very sharp battle up against other outlooks and methods, and other understandings of the problem and solution. So I wondered if you could talk a little bit more about that: the new synthesis up against other understandings of the world, of problem and solution.

AS: Well, I don't know that I can do justice to this in this interview, and I'm going to point people to more things that they could read on the website. People should go to the revcom.us website where they can find many of BA's works, and summaries of what the new synthesis is about. Also, people, including on the international level, who are interested in digging into this and understanding things on a more global plane, should definitely check out *Communism: The Beginning of a New Stage, A Manifesto from the Revolutionary Communist Party, USA*, and the whole section that is there about the new synthesis. And I have to say, once again, that people really need to get into the polemics in support of the new synthesis—such as the polemics from the OCR in Mexico and others who are defending the new synthesis and promoting it in opposition to a lot of different wrong lines in the international communist movement. These polemics are actually very rich in lessons and analyses, they are a valuable resource for people to turn to. And the journal of theoretical struggle, *Demarcations*, which can also be linked to through the revcom.us website, is a very important resource which is promoting broad engagement and struggle internationally over key questions, including analyses of things like what happened in Nepal, how and why the revolution in Nepal went off track, and other related questions.

I've spoken a little bit earlier about some of **the epistemological errors that have plagued the international communist movement**. Two basic types of errors. First, there is a kind of brittle dogmatic tendency among some organizations and parties—you can't talk to them about things further evolving in the science of communism, almost as if there were never any new objective developments and never anything new to be learned. They approach communism as if it were a bunch of sacred tenets, or like a catechism, rather than a living science. In fact, they refuse to approach it as a science, and to recognize that there's a need for a science itself to evolve, for new things, new understandings to be incorporated, as part of the accumulating knowledge needed to advance things. They act like they are frozen in time, and they do a lot of damage because of that. And then there's the other tendency which is also very prevalent, and very damaging in the world, and that represents the opposite error. There are some organizations and parties that are trying to evaluate things like why the socialist revolutions got turned back, why capitalism got restored, first in the Soviet Union and later in China, but who don't go about digging into this in a rigorous and systematic scientific manner. So they go very much off track, and they think they've made some great new discoveries, found a great new alternative, when all they are really doing is arguing for nothing more than standard bourgeois democracy. They tend to think that the basic problem is that everything has to become essentially more elastic, or that the problem in the old communist parties was just too much bureaucracy, and they just don't recognize or correctly dig into the actual epistemological and ideological errors, the basic errors of method and approach, that were involved in a lot of these situations—they don't sort it out correctly.

And if you want to see, by negative example, what happens when you don't apply the kind of rigorous scientific, consistently scientific methodology that Bob Avakian applies, then look at some of these parties and organizations that have gone off track in recent times, because they've based themselves only on very partial and unsystematic experience and understanding, and tried, frankly, to rely also on pandering to different social strata. They've undone some of their own best work and rendered worthless some of the sacrifices of their own people, because they haven't been consistent, and they've allowed themselves to be buffeted about by one pull or another, usually of bourgeois democracy, of those kinds of illusions. Sometimes you think that really what they want is the kind of bourgeois revolution of 1776 in the United States or 1789 in France. In the bourgeois revolution, when the bourgeois class came to power, it did bring in some bourgeois democracy and some liberties and some emphasis on individual rights that had not existed in prior feudal societies. But you get the feeling that some of these revolutionaries in today's international communist movement can hardly see beyond that kind of early bourgeois framework! They're also not keeping up with the big objective changes that have gone on in the world and in the class configurations in the different countries. And they're also frankly not proceeding from a very inspiring vision of how society should be organized, or how the struggle for a new society should be conceived of and undertaken as part of the larger international struggle, the international communist struggle to emancipate all of humanity.

That's the thing. You want to see where people go off track? Time and time again, you'll see that people are not proceeding back from the goal of emancipating all of humanity. They fall into the trap of promoting and trying to advance just one little corner of the struggle—in one particular country, or addressing just one particular set of needs. They're not proceeding from the most inspiring global vision of all this. And this leads them to go off track. There are many different ways people can go off track. Again, things like the theoretical journal, *Demarcations*, and the polemics in support of the new synthesis, and the *Manifesto of the Revolutionary Communist Party* that speaks to the end of a stage and the beginning of a new stage in the communist movement—these are all good places to start digging into these important questions. These are not easy things to get into if you're brand new to things, but they are things that people should be struggling around. And I think you should be able to pretty quickly see that, if there were more systematic engagement and deep discussion, not just of revolutionary practice in a tactical sense, but of these deeper strategic questions, in relation to those deeper strategic goals, such as what is argued for in the new synthesis, taking into account the way the world has actually been changing, and taking into account the breakthrough advances in philosophy and epistemological method that BA has been representing and concentrating, then I think there would be a much richer mix and much richer basis on a world scale for people to connect in a more unified way, in terms of a revolutionary direction, going in the direction of a communist world.

Great Upheavals in the World, and the Great Need for the Scientific Approach of the New Synthesis

Question: Looking on a global scale in the last few years, and then zooming back a bit to look at the last couple of decades, there's been a lot of places in the world where there has been major upheaval, and even revolutionary struggles—or, looking at the last few years, it hasn't been so much revolutionary struggles but there have been things like the initial upsurge in Egypt where the head of the government, Mubarak, was forced to step down, and other struggles that were part of the initial stages of the "Arab Spring." What I'm trying to get at is, imagine the difference, right in that context, if there actually had been a core of fighters for the new synthesis, and the new synthesis were really a force on the map globally, internationally.

AS: Yes, I think Egypt is a good example, and BA had a very good and important statement on Egypt that would be worth going back to and reading. You know, it's admirable when people stand up to fight against oppression...and people...in particular there were a lot of youth, a lot of college students in Egypt who really wanted to fight against an oppressive regime. Many of them want a better world, not just for themselves as individuals but for society generally, and they wanted to fight against the abuses. And you have many such situations that come up in the world where people are very brave and they put themselves on the line, and they sacrifice, they get jailed, they get killed—you're not talking about people who don't put a lot on the line. But if you don't have a deep enough understanding, not only of the **source of the problems** (in terms of global capitalism-imperialism and how it interpenetrates with the problems in a particular country and regime), but also of **what the direction of things has to be, strategically** so as to be able to work towards the kind of revolution that really would be emancipatory and that could bring in a new kind of system, a new socialist society to replace the oppressive society you're trying to overturn—if you don't have that kind of understanding, then you'll only get so far. You'll sacrifice a lot, you'll see a lot of brave people step forward and fight really hard.

But then, even if they experience some short-term successes, they'll get turned around. That's what happened in Egypt. The military came in, in a very heavy-handed way. So if you don't have that deeper and more strategic scientific understanding of both problem and solution, so to speak, you basically don't end up advancing very far, you tend to get crushed and pushed back, or you get stalled. We see that happening time and time again, so we have to figure out the ways to break out of that.

People have all sorts of confusion. People don't even realize there are no actual socialist countries in the world today. They think that China is communist, or maybe Cuba is communist. They're not communist. They keep the name, but there's nothing communist about their systems. Or people think that something like Chavez in Venezuela was some great new form of revolution. No, it isn't. They have never broken with the imperialist relations. There have been some genuinely good people fighting for, and actually dreaming of, better worlds, but if you don't make a materialist analysis, in a systematic and rigorous manner, then you don't really understand deeply enough what you're up against and how to break through. So, no matter what your intentions are, as individuals or as groupings of people, you are going to go off track, or you are going to get turned around, or you're going to get crushed. And that's what we have to avoid. Still today some people are confused about Cuba. Cuba never really had a genuine socialist revolution, they quickly became dependent on the Soviet Union, which was itself capitalist-imperialist by that time, at the end of the 1950s. There were elements of a revolution...there was bravery, there was sacrifice, there were some visions, some dreams on the part of many Cubans for a new world—but they didn't get there.

The Vietnamese people, during the war with the United States...first the war with France, then with the United States...the Vietnamese people made huge sacrifices...**millions** of Vietnamese people died getting rid of the occupation by imperialism, millions of people sacrificed, fought hard and died, trying to free their country from that kind of domination. And many of them were even looking for something more than that, something beyond just national liberation. They had some idea that they would create a better society. But they never even got a chance to start to build a new socialist society, because the people leading them did not have the right perspective and the right methods and were not really proceeding from that grander objective, and also because the broad numbers of people who were being led didn't understand enough themselves, they didn't have those scientific tools to actually be able to tell clearly enough what was going on and to be able to bring forward new and better leadership. So you had a lot of fighting and sacrificing on the part of the people, but then it didn't go over to a new place, to a better place.

These are some very bitter lessons, that we have to learn from. This is part of what I mean when I say that if you're a good scientist, **you learn from the mistakes**, from the missteps, from the misdirections. You **have to** learn from them. You can't just step over that. You can't say, "Oh well, that was a good try, let's move on, let's see what the next upsurge might bring." No. You really have to work to understand **why** things go off track if you want to figure out how to move things forward in a good direction for humanity. And that's a lot of what BA works on. That's what he works on all the time.

Question: In the context of what you're talking about, what would it mean if, not only in this society but on a global scale, this new synthesis really had broad influence and more and more people were engaging it, but then at the same time, in relation to that, there were actually a core of people taking up and really becoming ardent fighters for this new synthesis?

AS: Well, for one thing, it would become what people sometimes call **"a pole of attraction"** in this society and in the world. It would become **a frame of reference**. People all around the world could be checking it out and learning about it and trying, on that basis, to develop the revolutionary process in their own parts of the world, and as part of the world movement. One of the most bitter examples of where this has been lacking is what's been happening in the Middle East, for decades now. Those countries in the Middle East have been pillaged and dominated for generations by foreign imperialists who, directly or through oppressive regimes they installed and supported, have distorted development and created huge social problems, and certainly prevented development from going in a good direction and meeting the needs of the people. So there's a tremendous amount of anger and resentment about all this in the Middle East today, and of course there's a tremendous amount of anger and resentment over the fact that the imperialists' only answer to the unresolvable problems that they themselves have created in the Middle East is to bomb the hell out of people, city after city, country after country. That's the only way they know how to deal with things: They've got problems, bomb the hell out of them! They've been causing immeasurable suffering and destruction throughout the Middle East.

And the imperialists have also been backing the Israeli state, the oppressive state of Israel that has been basically carrying out brutal genocidal policies towards the Palestinian people in the region for generations now. The state of Israel is propped up by the U.S. imperialists and other imperialists, and it's intolerable, completely intolerable, that this should keep going on. And it's a bitter irony, as Bob Avakian has pointed out, that it is a Jewish state that is carrying out these atrocities. As he has put it, **"After the Holocaust, the worst thing that has happened to Jewish people is the state of Israel."** And that's absolutely true. Because when you think about the genocide of the six million Jews who died at the hands of Hitler and the Nazis in the Holocaust in Europe during World War 2, it is very bitter to confront the fact that the Zionist state of Israel, which was created after WW2 by re-drawing the map, driving Palestinian people out of large parts of Palestine and seizing their lands, that this illegitimately established and unspeakably brutal Zionist state is continuing to this day to lead Jews, of all people—people who should know something about what it means to be subjected to genocide—to themselves commit genocidal atrocities against the Palestinian people!

It's very important to recognize that, as BA brought out in the Dialogue with Cornel West, there are two ways of summing up the historical experience of the Holocaust. One way, the correct way, is to say: "Never again!—we should never let that kind of thing happen to Jews, or to any other people, anywhere in the world, ever again." And the other way, the very wrong way, which unfortunately characterizes the Zionist state of Israel, is to say, "Because terrible things were done to our people, anything we do now is justified, so we're just going to take over this land as a homeland for ourselves, and we're going to drive out and subjugate the Palestinians who live there and any others who dare to oppose us, we're going to do things to our benefit only, no matter what the consequences to others."

Look, it's important not to blur over the difference between Judaism and Zionism. Judaism is an ancient religion and culture, but Zionism is a political movement and political ideology—and today it is an outright fascist political movement and ideology. The Zionist state of Israel, which did not even exist before World War 2, set itself up as a Jewish state, as a supposed refuge for persecuted Jews, and is today supported by many, maybe even most, Jews, both in the Middle East and around the world. But, still, it's important to remember that not all Jews are Zionists. In fact, historically, many Jews in different parts of the world, including in the U.S., have been radical Marxists and revolutionary communists. My sense is there are quite a few people in the RCP today whose family background is Jewish but who chose to become dedicated revolutionaries committed to working for the emancipation of **all** of humanity; they rightly despise the state of Israel and would have nothing to do with it except to expose its crimes and denounce and oppose it. Unfortunately, today there are not enough Jews who feel this way. But there are some. And their numbers really need to grow! So I was very glad to see recently that at least a few progressive Jews here and there have banded together and stepped forward to publicly denounce the criminal policies of the state of Israel, proclaiming for all to hear: "Not in our name!" We need to see a whole lot more of that, we need to see a whole lot more progressive Jews stepping forward to say to the state of Israel, and to its imperialist backers in the U.S. and elsewhere, "Don't you dare commit genocidal policies towards another people in our name, in the name of Jews, in the name of the Holocaust and of the sacrifices that were made during the Holocaust. Don't you dare!" We need to see a lot more of that kind of attitude.

But look at this whole situation in the Middle East. You have these endless imperialist wars and armies of occupation. You have this incredible destruction. You have all this distortion by these imperialist marauders that's been going on for generations. And what emerges in relation to all that? These Islamic fundamentalist nut-cases that are going around trying to institute brutal, backward, feudal rules and laws, and trying to accomplish this by perpetrating acts of incredible brutality—and yes, it is brutal, and it is disgusting, and there's no justifying it. It is, however, true that, even though all this brutality they carry out makes for some very graphic videos and is objectively horrific, the atrocities these Islamic fundamentalist extremists carry out don't even come close to the scope and scale of the brutality and atrocities committed by the imperialist system, including United States imperialism, all around the world, over hundreds of years and right up to this day. The crimes against humanity and the brutal atrocities committed by the U.S. and other imperialists over the years have been on a **historically unprecedented scale**. That's just a documented fact. So people should not forget that. In fact, anyone who doesn't realize this should make a point of looking into the actual violence, the well-documented atrocities, that have been committed by these imperialists over so many generations. This is something that Bob Avakian brought out in the course of the Dialogue that people can look into for themselves. You know, it's the whole immense scope and scale of it. Let's not forget who the biggest marauders and biggest criminals and the biggest purveyors of brutality all around the world are, and they're right here in the United States and in other imperialist countries, the leaders of those countries. And we should be very conscious of that.

But, that being said, it's absolutely disgusting if the only real organized response and resistance to all those imperialist depredations ends up being left in the hands of these fanatical Islamic fundamentalists who want to institute backward religious views and impose strict repressive and oppressive rules on all of society, including by reducing women to the most low form of property. I don't need to go on and on. I think it's pretty obvious that their vision for society is not one to aspire to, to say the least, and that they are trying to impose their views by brutalizing people in the most crude ways. One thing that BA has repeatedly stressed, which I really agree with, is that **these can't be the only two options**: to go with the imperialists and their so-called democracy, their **bourgeois** democracy, which is frankly itself built on tremendous and ongoing brutality and violence, or to go with the "alternative" posed by the nut-case

Islamic fundamentalists...what kind of choice would that be? We've got to be able to do better than that! Many people get confused because, on the surface, the bourgeois democracy of the imperialists in a country like the United States does extend a few freedoms and liberties, especially to certain more privileged strata in the population, but we can't allow ourselves to forget that it does so on the backs of the people of the world that are oppressed and on the backs of the people in this country who are oppressed. The few benefits that people might experience from bourgeois democracy are built on the blood and bones (literally) of people in this country and all around the world. So that's on the one hand. And on the other hand there are these Islamic fundamentalists who want to return to some kind of oppressive religious caliphates and impose those kinds of rules on all of society. These two brutal reactionary alternatives cannot be the only options for the people of the world. And they don't have to be. There is a basis for fighting against the depredations of global capitalism-imperialism on a much more enlightened basis, on a much more genuinely revolutionary basis, one which seeks to unite very large swaths of humanity in the fight to get rid of the capitalist-imperialist system and to actually bring into being a whole new kind of society which would benefit the majority of humanity and which would be going forward, not backwards, on a much more enlightened basis.

Let's Not Get Frightened By Terms Like Dictatorship of the Proletariat...We Live Under a *Bourgeois* Dictatorship Now

Question: One of the dominant trends that you were talking about a little while ago that the new synthesis of communism is fighting up against, both in the world overall but including even among many who call themselves revolutionaries or communists, is the idea you were mentioning that the best humanity can do is basically going back to 18th century visions of democracy, bourgeois democracy. I think that actually gets right to the question of the dictatorship of the proletariat, in other words, rule exercised in the interest of the formerly oppressed and exploited classes. There's the trend that you were mentioning of people just throwing that out altogether and saying we just need bourgeois democracy and there's no need for the dictatorship of the proletariat. Or there's the other trend you were mentioning, the opposite trend of people just clinging dogmatically to the past experience of the dictatorship of the proletariat, and saying it's just a matter of repeating that. And then there's the new synthesis, which has a different conception than either of those two things. In terms of the dictatorship of the proletariat, it's firmly upholding it, but then it's also talking about exercising that dictatorship of the proletariat in a different way than in the past. So I wondered if you could talk about the conception of the dictatorship of the proletariat with the new synthesis, and how that's going up against the dominant trends.

AS: Well, we've talked about some of that, but, look, sometimes people get frightened off by terms like dictatorship of the proletariat. Let's not mystify this, it's pretty simple, right? Until we get to full-out global communism, which transcends class divisions, the world is still going to be divided into classes, societies are still going to contain different classes, and the question becomes which class is going to rule and in whose interests. And the idea of the dictatorship of the proletariat is that the class of people—not the individual proletarians, but the **class** that actually has nothing to gain from the oppression and exploitation of others, that cannot even be free itself until all of humanity is actually free of oppression and exploitation—that class, and its fundamental interests in the fullest sense, should set the terms, should set the objectives. That doesn't mean that you poll individual proletarians and ask them to set policy—that's not how it works. And if you don't understand that, then people have to go into, again, what BA has brought forward to understand the **important difference between the concept of the proletariat as a class and proletarians, as individuals**. It's not the same thing. There are **objective interests** that correspond to a class formation in the world today that is not in a position to dominate and oppress others, and that actually itself cannot really be free until it breaks the chains for everybody. **That's the historical role of the proletarian class**, until the day when there are no class divisions at all, and there's just one overall human collaboration, when classes themselves and class divisions are transcended.

But what people also don't understand is that we live under a dictatorship **today**. It's **the dictatorship of the bourgeoisie**. I don't care if they like to call themselves "democratic" and talk about all their great freedoms and liberties. You actually look into it and see how it's set up and structured, and you'll see the whole thing is set up to exploit and dominate great swaths of humanity, both in this country and around the world, in order to be able to prop up and expand and consolidate this system and the rule of that tiny sliver of society that actually benefits from that exploitation and oppression, that has an economy and corresponding politics that are all geared to accumulating profit privately, as capital, as domination over and exploitation of others.

We can't get into a lot of this now, but I want to encourage people to study what is meant by **the fundamental contradiction of capitalism**, and to get into some of the things that have been written about this by Raymond Lotta as well as by Bob Avakian. That the way the world is set up, one way to think about it, is that there's a bitter irony in that the work that is done all over the world to provide for the material requirements of life—to produce the food, to build the housing, the shelter, to build roads and hospitals and provide medical care, the work that gets done everywhere in the world to meet human beings' requirements of life—is actually done primarily today through a lot of collectivized work, through a lot of **socialized production**, in other words with large numbers of people being organized to carry it out. You're not talking about a few dozen people. You're talking about billions of people on a world scale, and there's all this socialized production that's going on in an organized fashion, with people working together, being made to work together in order to produce these requirements of life. But **the appropriation** of all this, in other words, what happens to what gets produced, is privatized—it gets privately siphoned off, there's a **private accumulation** of the material surpluses generated by all that socialized production. And that contradiction is **the fundamental contradiction of capitalism, between socialized production and private appropriation**, and it generates a tremendous amount of anarchy on the side of the capitalists. So you have this tiny sliver of society, the capitalist class, that is appropriating the bulk of this socially produced wealth. It's privately appropriated, in the context of this insane system that is also characterized by continual competition between different blocs of capitalists, who are all pitted against each other and trying to advance their **own** agendas and their **own** profit-making machinery. And, on top of all that, the big decisions and policies about how to run society and what to do and promote and what not to do and promote, are made primarily on the basis of what's profitable and what's not profitable for these capitalist interests. So in this kind of system it ultimately doesn't matter, how much the people might or might not really <u>need</u> something. What gets produced; what gets allocated and how; how the society basically runs day to day—it all gets decided not on the basis of what the people need but on the basis of whether it's profitable or not for the capitalists, how it does or doesn't fit into their plans for out-competing each other, and so on. That's also how decisions get made about whether and when to plunder other countries, all while plundering sections of the people in their own country. And that's also why even in a really wealthy country like the U.S. you can have such extreme poverty and degradation.

You know, we could solve the worst of those problems pretty quickly. There's no reason for anybody to be homeless in a country like this, or really anywhere in the world, with the material resources that exist in the modern era. There's no reason for anybody to be hungry, or to be denied basic medical care. There is absolutely no reason. Human beings have actually gotten to the point in the history of humanity where the basic needs of the people, and then some, can actually be met, materially. There's no question of that. The thing that prevents that at this point in history is this fundamental contradiction, and this dominant anarchy, of the capitalist system. People need to check it out and dig into all this. And I know some of that is kind of hard, because most people aren't used to studying political economy, but people should work hard to try to understand at least the basics, including this especially important anarchy/organization contradiction that Bob Avakian and Raymond Lotta have written about, because it is actually a tremendous clue, not only to how the current capitalist-imperialist system works...how it is organized and to what ends... but it is also a tremendous clue to understanding the tension that exists within the current dominant system in the world—this capitalist-imperialist global system—which is currently truly straining at the seams because of its basic and decisive contradiction, which is preventing it from resolving any of the big problems of humanity. It cannot resolve those problems, even if it wanted to.

The capitalist-imperialists and their system cannot solve those big problems either in their home countries or on a global scale. That very contradiction is itself the basis for overthrowing this system and dismantling its institutions. Ultimately, this is what can bring it to an end, and bring something much better into being, **if** the people work on this consciously and work consciously to bring in a different kind of society, built up on a completely different foundation, which is geared to meeting the needs of the people in the broadest possible way, is motivated by that, and is organized in accordance with that. And the dictatorship of the proletariat is an instrument for doing **that**.

So, again, getting back to this question of "the end of a stage and the beginning of a new stage of communism"—this is not just some clever little phrase. It's either true or not true, and I believe the evidence shows that it's true. That there was a first stage of communism, of the communist revolution, that has ended. It ended with the defeat, the reversals, of the revolutions in Russia and then in China. When that first wave of communist revolutions and these first socialist societies advancing towards communism came to an end, and things got turned around, that was the end of a certain stage. There were no more socialist societies anywhere in the world. So, on the one hand, objectively, now you've got the opening, as well as the need, for a new stage, simply because that first phase has come to an end: Those first socialist revolutions made a lot of headway, and we can still learn a lot from what they accomplished and from the problems they encountered, but ultimately those revolutions were stopped from going any further forward and got turned around. That's what happened.

So now, objectively, you're starting a brand new phase, one where there are no socialist countries but this is still the direction to go in. But we can also go into it on a new basis, with the benefit of having learned a lot of critical lessons from the first wave, which Bob Avakian has helped to clarify and concentrate. You see, in my view, **the new stage of communism that is opening up springs from both objective and subjective factors**.

First of all, there are the significant objective changes that have taken place in the world in recent decades—obviously and first of all, there is the defeat of the first wave of socialist revolutions, which we've been talking about; but then there are also some other big changes that have taken place, such as changes in some of the relations and some of the class configurations in Third World countries (including the expansion of the urban middle classes and the related displacements and dislocations in the countrysides of those countries), and there have been some significant changes as well in the class configurations and societal mixes of the imperialist countries. The world keeps changing, right? All this needs ongoing analysis. But the point here is that there are significant objective changes that have taken place all throughout the world in recent decades that must be taken into account and encompassed—and the methods of the new synthesis actually help to do that, to analyze those changes, what they mean, and what their implications are for the next round of revolutions. So, that's one thing.

But then, and very importantly, there's also the **subjective** aspect of what is opening up a new stage of communism. A new stage in the development of the subjective factor, a new stage of communist science. It's a whole new level of theoretical breakthrough. What's involved is not just a few more incremental advances in communist theory and methods, but a genuinely qualitative leap—in my very strong opinion, BA's new synthesis of communism represents a qualitative leap in the development of the science itself. On at least the order of the kinds of leaps brought about in different ways by the likes of Marx, Lenin, or Mao at earlier historical junctures. The science keeps advancing, including through critical analysis of the accomplishments and shortcomings of past syntheses, through which it builds and advances even further, and this is simply the latest and most advanced synthesis of the science of communism yet achieved. This isn't just an opinion—I believe it is a pretty straightforwardly demonstrable fact. So that too—that forward leap in the development of the subjective factor—is part of what is ushering in a whole new stage of communism, whether most people realize it yet or not.

To be clear and to avoid confusion, it is not that the big changes in the objective conditions that I've mentioned were **necessarily, or somehow automatically**, going to be accompanied by the subjective changes. It's not a one-to-one thing. There could have been these major changes in the world, but no subjective change, no qualitative advance of communist science, no new synthesis. Or, conceivably, there could have been a further advance in the science of communism, a new synthesis, without all the objective changes that I've mentioned taking place. But the fact is that Bob Avakian has done the work, by digging deeply into these major changes that have occurred, and drawing from many different sources, learning from other experience—and the qualitative advance of communist science, the new synthesis, has resulted from all of that.

What Does It Mean, What Difference Can It Make, To Have a Party Organized on the Basis of the New Synthesis?

AS continues: Once again, think of the real difference it could make if the new synthesis were to spread, were to be broadly engaged with, throughout society, and were taken up by revolutionaries everywhere. Many of the revolutionary communists today are people who came out of the great upsurges of the 1960s, including Bob Avakian himself. This was a very rich period. But there's a tremendous need now for younger generations to take up this new synthesis, to work with it, to contribute to further advancing it, and to spreading it around the world. Again, I would use the example...besides the U.S. itself, I'll use the example of the Middle East. What a difference it would make if significant numbers of people, including young people in these Middle Eastern countries that are in such turmoil...if, instead of choosing between either promoting American-style democracy and aspiring to either move to America or to build up a similar system in their own country (with all the horrors that are involved with that), or joining in with these nut-case Islamic fundamentalists and all their horrible ways of trying to restructure society—if, instead of choosing one or the other of those no-good options, there were some significant blocs of people, including significant numbers of young people, who were delving into the new synthesis of communism, studying it, debating it, really grappling with it and figuring out how to apply it in the context of their own countries—this could provide a real alternative, a genuinely positive alternative. They would, of course, have to figure out what it means concretely to apply the new synthesis to the particular conditions of their particular countries and societies. But the key methods and principles of the new synthesis would apply anywhere. They could take that up, and it would provide a positive alternative to both those bad alternatives. It could become a rallying point in places of the world that are in turmoil, of which there are many.

Question: Continuing with the point you just made about the difference, the tremendous difference, it would make if younger generations took up this new synthesis, I did want to ask specifically what you think it means that there's this vanguard party, the Revolutionary Communist Party, led by BA, that bases itself on the new synthesis of communism that BA has brought forward, and the need for that party to grow and for people to join that party.

AS: Well, again, I would refer people to the website revcom.us, where there are some articles that get into **why** a vanguard party is needed. Why you can't make a revolution without one. I think people would get a lot out of digging into some of that. And your question is a good question, because I think it's a question that people don't discuss a lot, or not enough. How are you going to help make an actual revolution without being really disciplined and really organized into a revolutionary organization, in other words, a revolutionary party? It's not going to be enough just to function as atomized individuals or even to just get together with handfuls of like-minded individuals in a somewhat disorganized manner.

There's a statement on the revcom.us website, *Get Organized for an Actual Revolution.* If you understand what an actual revolution is, what it involves—that it actually does require getting to the point where you can dismantle the existing state apparatus and replace it with a completely different state apparatus, different organs of power, that you have to seize power and organize society on a new basis with new institutions—how are you going to do all that without a very tightly cohered and organized body of people, who are very committed and dedicated to doing that and who are willing to function in a very disciplined and organized way? Many people would probably recognize the need for tight and disciplined organization later on, when things get to the point of military struggle, or things like that—people think about disciplined armies, and so on. But what about for the current phase of things, where what's mainly involved is political struggle, **fighting the power primarily politically** for now, working to unite people on that basis, and working to transform the thinking of the people, but doing so in a way that will lay the basis for being able to "go for the whole thing," for the actual seizure of power, when the conditions exist for doing so? Even now, under the conditions of today, you'd better not just function in an individualistic way, or in a scattered way, like a bunch of disorganized individuals who sometimes work together and sometimes don't, and who are constantly pulling in different directions and end up undermining even their own best efforts. Making revolution is a complex multi-faceted process which needs to pull together many different components of the struggle and keep them all pretty much on track and advancing in a certain direction. So you'd better be as unified as possible, you'd better all be pulling in the same basic direction, and you'd better be recruiting more people and constantly expanding the ranks of the disciplined, organized body that can provide ideological and political strategic guidance and direction to ever broader people in society.

Question: And what does it mean to have a party that's based on the leadership of BA and this framework of the new synthesis?

AS: Well, a party is obviously made up of a lot of individual human beings, and not all of them see eye to eye on everything or understand things the same way or function all at the same level. And, as I said before, I think there's a tremendous "gap" between Bob Avakian and pretty much everyone else. He's like "miles ahead of even the best of the rest," as someone once said, in terms of people in the RCP as well as people outside the Party. That's just objective fact. But OK, we can work with that—first of all we can learn to more deeply value and appreciate what it is that BA has developed—that he has come to concentrate and that he is constantly modeling for others—which objectively puts him so far ahead of the rest of the pack, so to speak. We can do our best to learn from him, in particular by closely studying his whole method and approach to things. And we can work to at least significantly "narrow" the gap, in an ongoing way, including by having a good attitude about being led and learning from advanced leadership, and by actively contributing ourselves to continual grappling with the new synthesis and how to apply its key principles and methods in an ongoing way to further developing and advancing the movement for revolution.

People should understand better both what it means to be *willing to lead*, and what it means to be *willing to be led*. It should be a two-way street of mutual and inter-dependent responsibilities and the furthest thing from a passive or one-sided process. Being provided leadership, if it's good leadership, doesn't mean that you're just being bossed around or given orders all the time! [laughs] That's not leadership. Good leadership consists primarily in training people in overall orientation and method and approach, and in this way giving them the tools to contribute as much as possible themselves to the advance of the overall larger process and objectives, and to in turn train others to do the same.

And again, a revolutionary party has to function as a unified body, which is why there's a concept, **democratic centralism**, that people can read about in the *Constitution of the Revolutionary Communist Party*. Democratic centralism is not just a question of people following orders or being disciplined, although it is that, too, for instance in relation to things like carrying out assignments and responsibilities. But, democratic centralism involves much more than that. Democratic centralism is, most fundamentally, a scientific concept about **epistemological** discipline. It doesn't mean that people are slavish. But it means that when analyses and syntheses are developed at leading levels, and strategies and methods for a particular period of work and for prioritizing things are being developed, then the Party as a whole should function as a unified body to take this out to the best of their ability into the world. Like good scientists who are working in a coordinated and disciplined way on a scientific project. In this case, they're working on the project of transforming society, transforming the thinking of blocs of people, of fighting the power around egregious outrages, all in a disciplined and unified way along with the broadest numbers of people that can be united to do so, and doing all this in a coherent way. And then, if Party members have differences and don't agree with certain things, they have a responsibility to raise their questions or disagreements, in a systematic way, through the appropriate channels. This, too, is part of the scientific method and process.

You function in a united, unified way, but then internally people discuss and wrangle and debate and raise questions or disagreements and modifications, and so on, so that there is actually a genuinely collective process. You know, there's that formulation of the RCP, that the Party's collectivity is its strength. It is of course being given centralized guidance: Guidance is being provided regularly to the Party and to the people around the Party who are interested in learning about this guidance and orientation. So, yes, the Party is being guided, it is being led. It is being guided by BA, including through his works, and it is being guided through the website revcom.us, through Party documents, and so on. So there's definitely guidance, there's definitely leadership, being provided. At the same time, people are not—and should not be—passive. People in the Party, at every level, as well as people outside the Party, should definitely be raising their own thinking, their questions, their disagreements, but in a substantial way, and in an appropriate manner. In a manner that will likely contribute in a positive way to the overall process. This doesn't mean that you have to have a whole deep analysis of something before you can raise a question or possible disagreement, but whatever you raise should at least be **with the right spirit**. What I mean by saying that this should be raised in a substantial way and with the right spirit, is that it should not consist of a bunch of "nyaa-nyaa, crotchety-crotchety, complain-complain, I don't like this, I don't like that." You know what I mean? That doesn't get anybody anywhere. Even if it's a simple question or a simple disagreement, it should be raised in the spirit that we're all trying to get to a better world, and that's what we should all be doing together.

That's why, once again, I feel that in the Dialogue, Bob Avakian and Cornel West set a good example that other people should follow. They have some substantial disagreements, which they made clear. But they also identified substantial points of unity, and manifested a sort of joint moral conscience, in terms of fighting oppression. And so they could find the way to work together while still talking to each other and talking to the general public about what some of their differences are, and challenging people to grapple with that, not being afraid of seeing people grapple with that.

Question: So the Party enables people to collectively, in a unified way, apply the new synthesis of communism to reality, to grapple over that new synthesis and its application, and to further develop it.

AS: Right. Like a good team of scientists, with BA in the position of team leader, overall team leader, and with other people playing their roles to the best of their abilities, in accordance with their experience and understanding, and with the development of their ability to grasp and apply the scientific method. It's very much as if you were trying to solve a huge scientific problem in the natural sciences—for instance, if you were trying to find a vaccine for Ebola, or trying to cure cancer, or trying to figure out how to turn back global warming, or trying to stop the deforestation of rain forests—and, in order to increase your chances of succeeding, you set about organizing and unifying a whole bunch of scientists to work together, at different levels, with different abilities, different levels of experience, but all united in their willingness to: work coherently together, using the best possible scientific methods; study and build off of the accumulated knowledge in their field so far; bring their own creativity and initiative to bear; and follow the lead of a team leader, who is best able to provide overall guidance and direction for the project as a whole, and who has demonstrated, and models for others, an especially advanced and developed level of knowledge, expertise, and methods relevant to the particular field, and to the problem to be solved.

Well, in the "field" of applying scientific methods to "solving the problem" of emancipating all of humanity from the bone- and soul-crushing system of capitalism-imperialism, the person today who is best imbued with these qualities and most able to assume the responsibilities of team leader is clearly BA. Again, this isn't just my personal opinion—I believe this is a clearly demonstrable fact. There's simply nobody else today working at quite this level.

So we should consider ourselves lucky to be able to work with, to take guidance from, the person who happens to currently be "the most advanced expert in the field," and we should take full advantage of his overall guidance and leadership if we are serious about making revolution, in the right ways, and with a real chance of succeeding.

But everyone does need to pitch in. Look, you go out into the world and you're trying to transform material reality, you're trying to transform society, and of course sometimes you're not sure what you're doing, or you run into obstacles or you start going off track, or whatever. But you can learn from all that, too. Don't step over it. If you do go off track, or if you run into problems, don't just try to skirt it, ignore it, finesse it or just move on to the next thing. Instead, leave your ego out of it [laughs] and confront it, face it, figure it out. There are bound to be lots of problems and lots of mistakes made, and the problems you are having are probably shared by quite a few others. So let's just talk about it, let's collectively learn from it, in order to keep getting better at what we need to do.

And if, on the other hand, people are doing things that are making breakthroughs, are making advances, don't keep this to yourself either. Don't just think, "Oh, how cool!" and then keep it to yourself. **Report** on what you are encountering, on what you are learning out in society, on what is actually advancing things and connecting with things. Because there will be important insights and new experiences that come from every level, including from people at the base of the Party and from the people outside the Party who work closely with the Party. But knowledge of this needs to be shared. You don't want to squander any of that.

So, again, there's the responsibilities of leadership and the responsibilities of the led, at whatever level. The responsibility of leadership at every level is to lead. The responsibility of the led is to take leadership, to follow leadership, with the orientation of not being slavish but of fighting oppression and working towards the emancipation of humanity. And, in the course of taking leadership, learn to be a leader yourself and spread that leadership and that revolutionary consciousness and organization throughout society.

Question: So, with the Party there's a basis for this new synthesis to become a material force in the world in a way that wouldn't be possible without this Party.

AS: Yeah, without an organized party, without an organized revolutionary movement, it would just end up being small numbers of people talking to each other behind walls.

Question: Returning to the work and the leadership of Bob Avakian, and the role he plays in the world, as you have said, this is very contended. There are some people who really love Bob Avakian and what he's brought forward and represents and the role he plays in the world, and there are some people who really don't like this. And I wondered if you could get into that further.

AS: I think that's actually a very important thing to dig into more deeply, because there's a lot you can learn from digging into **the reasons why** so many people do love Bob Avakian and his work, and, at the same time, **the reasons why** so many people hate Bob Avakian and his work—or at least hate Bob Avakian, because, again, many of the "haters" hate him without really knowing his work—they typically don't really study his work, they don't really get into the specific arguments, they don't really engage the analyses and the syntheses, they don't come up with serious, substantial criticisms. What prevails, at least these days, among most of those haters is more in the nature of petty slanders, insults and personal attacks. It's very low-level, low-minded kinds of attacks, and there's a real shortage among most of those haters of any kind of substantial analyses of the societal problems that are being tackled and the solutions that are being proposed. With a few exceptions, you don't see people writing papers or giving speeches that are really engaging what Bob Avakian is saying about the strategy for revolution, how to develop a revolutionary movement in the United States, why revolution is necessary and possible, how we could have a realistic chance of winning, what kind of society we could build up, and just how would we go about it. You know, there's a whole body of work that Bob Avakian has developed, over decades, with very substantial documents and analyses of these questions, and he's done a tremendous amount of work to make this readily available. And yet these haters are not so much, in this period at least, characterized by people who really develop counter-arguments and substantial counter-analyses. It really is much more gutter talk and snark. And this has something to do with the prevailing culture. There are many people in the culture generally these days who seem to make it a hobby to tear down other people with petty slanders and insults. It's all over the internet and stuff. But, with regard to Bob Avakian specifically, this takes the form of a tremendous amount of passionate vitriol against him. And you have to ask yourself: Why would some people so passionately hate someone who has spent his whole life dedicating himself to trying to serve the people, and to the emancipation of humanity? You can agree or disagree with his specific arguments and analyses, you can have substantial differences, and so on, and you can debate these and discuss these in a principled manner. But why on earth would you be trying to personally attack and tear down someone who has not been trying to promote himself or sell you anything or feather his own nest, or anything else of that nature? Quite the contrary, he's dedicated his entire life to serving the people and trying to come up with solutions to the horrors of the system and to being able to bring into being a new society that would be better for the vast majority of people in this country and the world. So why would anyone actually have such passionate hatred for someone like this?

And it's important to make a scientific assessment of those kinds of tendencies, besides just recognizing the prevailing culture of snark, which is a disgusting feature in society more generally these days. Again, I feel you have to further explore **why** some of these haters, most of whom today don't even bother to familiarize themselves with BA's extensive body of work or engage it with any seriousness, are nevertheless so bent on spewing so much hateful vitriol in his direction. Why is that, really? And I think that, to get at what's really going on with this, you have to ask those people some pointed questions: What's **YOUR** analysis of the problem? What's **YOUR** analysis of the solution? What are **YOU** putting forward, and arguing for? What kind of resistance are **YOU** organizing? What are **YOUR** strategic objectives? What kind of new society are **YOU** proposing and how are you proposing to get there? If you don't think this system needs to be overthrown and dismantled through revolution, then what program and solutions are **YOU** proposing? What is **YOUR** plan for getting rid of the incessant outrages and abuses generated by this system and built into its foundations, such as the police murders of Black and Brown people and the slow genocide of mass incarceration; the patriarchal culture of rape, degradation and dehumanization of women and denial of the right to abortion; the wars of empire, armies of occupation and crimes against humanity perpetrated on a regular basis by imperialism; the closing and militarization of borders and brutalization and dehumanization of immigrants; the accelerating and multi-faceted degradation of the global environment that is being driven by imperialism towards a literal tipping point of no return. What is **YOUR** solution to all this? What do **YOU** propose?

We should be confronting those haters with such questions. We shouldn't let them get away with spewing hatred to try to tear down and diminish BA, and by extension everyone working with BA, just because they themselves have nothing much of substance or value to propose. If they don't like what BA and the RCP are analyzing and proposing, why don't they just go do their own work on solving the problems of humanity!

I think some of these people just want to keep one foot in the system, you know? Why are they kicking and screaming at the prospect of going towards a new society that could benefit the vast majority of people? Would they actually prefer to keep things as they are? This is particularly characteristic of some of the petit bourgeois strata, in other words, the people in the middle classes. Not all of them, of course, but some of the people in the middle strata want to keep at least one foot...Look, by definition, that's what **the petite bourgeoisie** is, right? It's the class that sits **in between** the proletariat and the most oppressed at the bottom of society, on the one hand, and the ruling bourgeois, the ruling capitalists, on the other hand. So they're kind of **in an in-between limbo**, and it's pretty common for many of them to hedge their bets and try to keep one foot in both worlds—one foot in the current system, because, if they're being honest, they still kinda like living under this system, from which they still derive quite a few advantages and privileges; and one foot which, at least in their better moments, might be willing to step into the future, because many of them do recognize that this system is a horror for the people at the bottom especially, and many, again in their better moments, would sincerely like to get to a more just and equitable society. But they are often reluctant to upset the applecart and do what needs to be done to get there. So they remain torn. Some of them end up playing very positive roles and contributing in various ways to the overall process aimed at emancipating the oppressed, the exploited, and ultimately all of humanity. But some of them get downright nasty and try to hold back, and tear down, those people and forces that are actually going forward and working on getting organized for an actual revolution and a fundamental change in the system running society. So judge for yourself.

And we can talk about that some more. But, I guess I'd like to ask people to think for a minute about respect and about disrespect; about people who prove, over and over again, that they have principle and integrity, and a generous and broad-minded spirit, and who are trying to change the world for the betterment of humanity, versus, on the other hand, people who seem to spend a great deal of their time mainly tearing other people down, and spreading petty, snarky, vindictive slanders and insults and launching personal attacks while themselves having very little to offer people in terms of a viable and realistic way out of the horrors of the system, and very little to offer people in terms of a concrete plan for how to remake an entire society on a basis free of institutionalized exploitation and oppression. So, please, people, **think about this contrast**. Because it **is** burdensome and damaging when there are people who are always kind of nipping at your heels, trying to get in the way, and especially trying to get in between Bob Avakian and the people he's trying to speak to—constantly nipping, nipping, back-biting, trying to tear down. Is this really what should be going on?

Have some principle, have some integrity. If you have disagreements on matters of substance, by all means write them up, make speeches, make analyses, make them known. If you have alternative programs and approaches, by all means bring them forward. But do it in a principled manner, with principle and integrity. Don't go down in the gutter, nipping at people's heels and trying to get in the way, trying to prevent them from connecting to the people they are trying to reach.

Some Thank Yous That Need To Be Said Aloud

AS continues: Especially in the face of not just the hardship and difficulties, but also the slander, the snark and gutter attacks that some people never tire of spewing forth, I'd like to say some thank yous, because I think there are some thank yous that good people have in mind sometimes, but that are not enough, not often enough, said aloud. So let me say some of those thank yous aloud right now.

Thank you, first of all, to Bob Avakian, for his tireless dedication and many personal sacrifices over many decades. Again, all he's done his whole life is work tirelessly to serve the people, not for personal advantage or to feather his own nest. Thank you for never giving up, for never selling out, for always trying to more deeply understand the deep root causes of the great unnecessary suffering experienced by so many here and around the world. Thank you for buckling down and doing the hard work to apply consistently scientific methods to uncover the truth of things, wherever it might lead, however uncomfortable it might be, and then following through to bring to the fore "the logic of the logic"—that a revolution is not only desirable, but absolutely necessary and also possible. Thank you for your generosity of spirit and your broad-minded inclusive and optimistic vision. Thank you for all your work in developing the vision, the strategy, and the concrete plans to advance towards the emancipation of humanity from capitalist-imperialist oppression, and then working tirelessly to spread this understanding and this strategy and this plan broadly among the people—to thousands, to millions, to any who would listen, especially among the most oppressed at the bottom of society that so many in society would just feel comfortable throwing away, while you invite in all others who are willing to join in the movement of resistance and revolution. Thank you for telling it like it is, for doing systematic, scientific work on the problems, for giving of this knowledge and of yourself to all who would listen.

Thank you, also, to all the other comrades, the followers of BA's new synthesis of communism who contribute daily to this process, to the best of their abilities, and also often at great personal cost. Thank you for not giving up, for fighting through the exhaustion and discouragement, for dedicating your lives to serving the people, for striving to always learn more and contribute more, and on an ever higher level.

Thank you also to all those in the broader society who in many different ways donate their time, their money, their ideas, their legal expertise, their research, their organizing skills, their music, designs, paintings and other art works. To all those who open their doors and their hearts to welcome and assist the resisters and revolutionaries, thank you. To all those who have refused to bow down to social pressure, to turn their backs on the revolutionary communists, to shun or slander them, thank you. Thank you to those brave elements who have stood up in places like Ferguson, in defiant resistance, who are serving notice on the system that they will not take it any more, and who are working to put aside their own conflicts and the differences among themselves in order to stand up together to the greater enemy—this system and its enforcers. You inspire and motivate many, many more in this country and around the world. And you are being heard. Thank you.

Thank you to all the heart-broken ones who have suffered unimaginable loss and grief as their children and other loved ones have been brutally slaughtered by the police and other enforcers. Your cries of agony echo forevermore in the minds of the revolutionaries, and are constant reminders of the need to persevere to put an end to this horrible system. Thank you for standing up in the midst of your pain, and joining with others to fight and resist these outrages, to demand justice, to demand that these outrages stop once and for all, so that no other family should experience ever again such needless pain. What you are doing is a fitting tribute to your lost loved ones, and will give strength to the movements of resistance and revolution which are working to get beyond all of this. Thank you.

And once again, coming back around full circle, thank you to Bob Avakian for the dream, the vision, and the ability to turn all this into concrete plans and a concrete strategy for the emancipation of all the oppressed and exploited, and all of humanity, and for envisioning and mapping out how things really could be so different, and so much better, for the vast majority of people on this planet. Thank you for your willingness to shoulder the responsibility to lead. Thank you.

Question: Well, that was really, really deep, and heart felt, and really right on. I actually hope people reading this will take a minute, more than a minute, to think about what you just said, and to really reflect on it. There's a lot to learn, both from what you said, but also the whole spirit in which you said it, because, as you were alluding to, there's this whole culture of really deep cynicism and snark, where it's considered "uncool" to express very earnest and deep appreciation of anything actually in this culture. It's considered very uncool to express sincerity and earnestness, particularly when it has to do with changing the whole world. As you spoke to, there's this whole culture of snark and pettiness and nastiness. So, mainly, I want to let what you just said speak for itself, because it was very profound and very much to the point, your thank yous and your appreciation overall, the whole of what you said, and in particular your appreciation of BA and what you were pointing to about how this is someone who has dedicated his entire life, who has spent decades, in an unwavering way, working on how to emancipate all of humanity, bringing forward the science and the theory for that, and leading the whole movement and a Party to make revolution to do that, never giving up and going deeper and deeper into that process. So, again, I really would encourage people reading this to reflect on what you just said, and the whole way in which you said it.

Why Is There So Much Cynicism and Snark—And How Can That Be Transformed?

Question: One point you were making that I wanted to pick up on and go further with, is the contrast you were drawing between principled disagreements and serious, substantive engagement with the content of BA's work and what he's brought forward, and on that basis, people putting forward what they agree with, what they don't agree with, and their questions, versus this nasty, in-the-gutter garbage, which really is just outrageous. I don't want to dignify it by actually getting into detail, but a lot of it is just absurd and nasty and really mean-spirited. And what would make anybody think that that would be OK, these kinds of lies and personal *ad hominem* attacks, and slander? As you said, there's something really wrong with the culture when that kind of stuff is considered OK, considered acceptable. And that is especially so when it's directed against somebody who is leading a revolution and working on the emancipation of all of humanity. So, I really agree with what you're saying about the viciousness of that. And you contrast that, for example, with what we see in the Dialogue between Bob Avakian and Cornel West—and it's very clear that Cornel West has disagreements with Bob Avakian, Cornel West is not a revolutionary communist. So, in addition to a lot of unity, they also have some sharp disagreements, but it was really striking, as you were saying earlier, to watch the incredible warmth and mutual respect and love between them, even the way they were hugging during the Dialogue. And, if you watch the live stream of the Dialogue, you can see at the end that Cornel says to BA, I love you, and BA says, I love you, too. There's a really great spirit of mutual respect and love there, and we need a lot more of that in the culture. And there's your proof right there that people can have very sharp disagreements but still relate to each other in a principled way, getting into matters of substance, and not tearing each other down. But then you contrast this with what you were talking about, this gutter shit, this just really nasty shit, and it's like: How is this allowed to fly? This is ridiculous, this is outrageous, and has nothing to do with even trying to change the world when people are indulging in that kind of nastiness.

So now I'm ranting a bit. But to focus back here, part of what I wanted to ask you is to get a little bit more into this question of why do people engage in this—this nastiness, this snark and slander? Where does it come from? Why do people do that? I think this is something that sometimes confuses and disorients people, including newer people coming forward. They get introduced to the revolution, they get introduced to BA, they feel really fired up, they expect, when they go and talk to people in society about it, that everybody's gonna love it, and then they kinda get thrown off. Even if they don't agree with this kind of nastiness, and even if they themselves don't like it, they are kind of confused, like "what's going on here, this is about changing the whole world, people should love this, anybody who's calling themselves a progressive should love this." So, again, the question is: Where does this snark and slander come from, and why does this exist, what does it represent?

AS: We've talked about this some, but there are a number of factors that are worth analyzing more fully. One is the times we live in now, and the degree to which this culture of snark is prevalent in the culture more generally. The internet and social media in particular is full of personal attacks and disgusting slanders against all sorts of people, tearing down all sorts of people, in particular celebrities and people in the public eye, people who play a public role. Some of this has to do, I think, with the post-1960s decades, during which, basically, the revolutionary movements have ebbed. So I think this culture of snark is in some ways connected to the reversals of the revolutions and of the revolutionary movements around the world, and to the loss of the first wave of the socialist project. In other words, the reason I'm connecting it to that, is because there was a time, in the 1960s, in the movements of that period, where it was certainly the case that people had all sorts of differences—there were all sorts of groups, and people would have all sorts of disagreements and wage all kinds of polemics. But overall the prevailing culture and ethos was very different. Sure, there was some snark and disgusting petty stuff back then also, there was some opportunism...especially coming from people who didn't have that much to offer, and who kind of made a career out of tearing down other people. But it wasn't anything like the snark culture of today. In general there was much more hopefulness, much more optimism about the idea that you could actually change the world for the better, and that humanity could eventually get to a whole better place, a more generous place, a more collaborative and communal kind of place. And whether they were communists or not, a great many people back then really were aspiring to something better than the dog-eat-dog world of the capitalists. And it seemed real, it seemed possible, it seemed obtainable. But then with the defeats, in particular of the revolution in China, and the setbacks in revolutionary movements throughout the world since that time, the waning of the revolutionary movements, coupled with a very intense ideological counter-offensive on the part of the capitalist-imperialist system...

the whole way in which they've very systematically promoted a lot of anti-communism, encouraged the publication of sob story narratives about the Cultural Revolution in China, and in other ways promoted portrayals of communism as a horror and communist leaders as monsters...the whole mood and culture started to shift. Look, this ideological counter-offensive aimed at disparaging and distorting revolution and communism, it's been going on for decades now, and, yes, it **has** influenced people who once knew better. A lot of people, to the extent that they've encountered that and swallowed it whole and uncritically, without really looking into the facts, have tended to turn away from the more optimistic outlook many of them once had.

A lot of the "'60s people" have ended up becoming very cynical. A lot of them had initially tended to romanticize the basic masses, imagining that the most oppressed in society must necessarily have the noblest qualities and be some kind of saints, or something; and when they discovered that in reality people are complex, that they lead complicated lives, and that they're not all saints, this disoriented a lot of the initially progressive, or even revolutionary, middle strata people in particular. They didn't apply scientific methods, they didn't examine things with enough scientific materialism. So they could no longer see beauty in the basic people, the potential in the people. There was that famous movie, *The Big Chill*, which represented a kind of turning point in the culture. It presented a bunch of '60s liberals, who had once been part of the youth and college student progressive or radical counter-culture, basically becoming disillusioned in later years, turning against the people at the bottom of society and becoming disgusted with them, starting to see them as just a bunch of guilty criminals and degenerates. That's what a lack of science can do to you! They flipped from overly romanticizing or idealizing the basic masses as some kind of angels, to reducing them to faceless criminals devoid of humanity, one could only fear and despise.

This kind of thinking has been a problem for decades—there's been a sort of unhealthy atmosphere that way. And then the internet ended up providing a broader platform for spreading and egging on a lot of cynicism and virtual bloodlust. It has provided a mechanism for engaging in a lot of tear-down culture in general, and for doing it at a safe distance, often anonymously, without really having to be responsible or accountable for any damage this might cause. It's become a thing where people can hide behind anonymity in cyberspace and just spend a lot of time in their pajamas making snide and cynical remarks, spewing negativity and tearing down other people, and that's what a lot of people do. People who spend their time doing this reveal their nasty disposition, their shortage of imagination, their lack of moral conscience and social responsibility. And, on the part of people who once knew better, it represents giving up on actually even **trying** to make a better world. So, I think that's part of what's happened.

A Materialist Understanding—A Material Basis for Why Some People Have Trouble with the Prospect of an Actual Revolution

AS continues: But then there are some other components. I mean, when it comes to Bob Avakian, there's still a major problem that a lot of people simply have not yet encountered his work. And among people who have some integrity and principle, whatever their background is, when they seriously get into it—they watch a film, they listen to a speech, they read a book, or whatever—they're more likely to be impressed than not. They may have questions or disagreements, but they see there's something serious there that's worth grappling with. But the attackers, the haters, often don't even begin to engage the work.

And why is that? Well, as one important part of it, you have to look at how a society is organized into **social classes**. When we get to a full-out communist society, around the globe, for the whole planet, we'll be transcending all these class divisions, we'll be getting beyond that. But until then, societies are still going to be divided into classes. In a society like the U.S., you have a small sliver of people at the very top, the capitalist-imperialists who run this system, and they run it to their advantage. And they run it by controlling all the major institutions of society, including the government institutions. That's why **elections are such a farce**, you know. There is no such thing as genuinely free elections in a society like this. The candidates are selected according to how much they'll play the game, basically, in favor of the capitalists. Even if they have differences among themselves, they're all basically on the same page from **that** standpoint. **The capitalists**, they control the big institutions, they control the government, they control the courts, they control the police and the armed forces, all the enforcers of this system. So they're a small but obviously a very powerful and dominant class in this society.

At the bottom of the society you have a class of people, called **the proletariat**, the people who don't own any means of production—land, factories, and so on—and who just have their ability to work, and have to sell that to the capitalists in order to live. In the United States this includes many poor whites, not just Black and Brown people. But it does include very large numbers of Black people, and Latinos, and other oppressed nationalities. That's actually a large group of people. There are many who are unemployed, some who are more or less permanently unemployed, as well as others who are more regularly employed. And there are also large numbers of immigrants coming in from Mexico and other countries, people who have been driven by the workings of imperialism—by the devastation, or the distortion, of their countries by imperialism—to seek work and livelihood by coming to the United States. They too are, by and large, part of the proletariat and the oppressed at the bottom of society. And whether individuals in all those strata at the bottom know it or not, whether they agree with it or not at any given time, their **objective interests** actually lie in the kind of complete radical upheaval and transformation of society which can only be accomplished through an actual revolution—a revolution to lay the basis for a completely different kind of society, one that would no longer allow the kind of exploitation and oppression that most of them experience and suffer under.

But then in a country like the U.S.—and this is where I'm gonna get back to the snark culture again—in a country like the U.S., you also have a very large **middle class**, the layers of people that sort of stand "in between" the people at the bottom and the small sliver at the top that runs everything. These middle class people are what Marxists call the petite bourgeoisie, because it's a small bourgeoisie, it's not as powerful as the big bourgeoisie that runs things. **These middle strata don't really run anything.** They might **think** they run something, they might think that they're free to influence things, but they really have no power themselves. But they are nevertheless occupying a position in society that's quite a few notches above the people at the bottom. They generally enjoy a relatively more privileged existence, even if there are problems in their lives.

Of course, they're not all one uniform class. This is true of **any** class. There are always many different individuals, with different perspectives, in any class. So you might be from the most oppressed and be a total jerk, and you might have been born into a wealthy ruling class family and be a very enlightened person. The particular circumstances an individual is born into doesn't in itself determine how that individual is gonna end up thinking and acting. So when I'm talking about classes here I am talking about more general patterns. But among these people in the middle strata, the petite bourgeoisie, it's a fact that they occupy this in-between position in society on a material level, and many of them, especially among the more educated, enlightened types, the so-called liberals, or progressives, or whatever, there are quite a few of them who actually are disgusted by a lot of the outrages in society, the injustices. They might take up the fight against mass incarceration and police brutality, or they might fight for abortion rights, or they might denounce the imperialist wars in the Middle East, and many of them are very concerned about the despoliation, the degradation of the global environment, they're very worried about what's going to happen with that.

So there are quite a few middle class people who actually **do** have deep criticisms of the way things are.

But "not liking the way things are" is not the same thing as having an understanding of the need for radical change, or a desire to go there. And these people, they're often kind of schizophrenic, you know. Some of them might say that they **want** radical change, or some might even say they want some kind of **revolution**. But then, when it comes right down to it, they're kind of stuck in that in-between place, because they do enjoy some privileges in the current society. For instance, they might enjoy a pretty decent standard of living, and be able to more or less meet the needs of their children and other family members. They might be able to take vacations and live in a decent house, and they might even like their jobs and get a good salary. So, do they *really* want to rock the boat? The point is that there's a certain order and stability to their lives, even if they genuinely and sincerely hate some of the abuses and outrages that the capitalists perpetrate on the people. So, their better natures, if you want to put it that way, their better sides, aspire to something better in society.

Buuuuuttttt...[laughs]...maybe, in their daily lives, it's not all that bad for them. So when they think of radical change and revolution, they start to think of chaos, they start to think of suffering, they start to think of people dying. They start to get all twisted up about the question of violence, of revolutionary violence. Look, this is not the issue for today in this country, but obviously, a real revolution is not a dinner party, as Mao has said. When you get to an actual revolution, yes, there will be battles; yes, people will die; yes, people will suffer.

But see, the thing is, if you have the perspective of the oppressed, of the people at the bottom of society, there's a lot of violence that's *already* going on, violence that the people are *regularly* subjected to. The blood is already running down the streets. The bones are already being crushed. The spirits are already being suffocated. On a daily basis, and on a massive scale, both here and around the world. This is the daily reality for the people at the bottom. So, you know, for the people at the bottom, how much chaos and disorder and societal convulsion will occur at the time of a revolution—or, for that matter, in response to movements of resistance, protests and things like that—that tends to be less of a question for them, less of a preoccupation. Because of what they *already* experience, so regularly, in day-to-day life under this system. People at the bottom **do** tend to have real deep questions about whether or not we can really **win**. Individuals will be brave and make sacrifices, but nobody wants to fight and sacrifice just to end up being crushed and getting nowhere. So is there a realistic chance of winning? What would that look like? What would the new society be like? Who are our allies going to be? Who can lead us? Will we end up being betrayed? These are some of the very real and serious questions that people at the bottom of society are more inclined to want to raise and discuss.

But those people in the middle strata, they're more like "ehhhh...sure...we kinda want a better world...we kinda want big changes...but...uhhhh...we also kinda **don't**!" [laughs] They're in the middle. And some of them do end up rising above that confusion and schizophrenic aspect to actually dedicate their own lives to serving the people. There are many revolutionaries and communists who have come from those strata, especially among the educated students and intellectuals, and so on, and the revolution could not happen without them, without their roles and contributions.

But, on the other side of things, there are people who dig in, people who, in sort of an abstract way, might like to **talk** about radical change, or even revolution, but who actually hate and fear the real prospect of an actual revolution. They have kind of dug out a little space for themselves, carved out a little niche for themselves as just some kind of perennial "critics," as some kind of pretty tame "loyal opposition," always operating well *within* the confines of the current system, perhaps willing to criticize it to some extent, but unwilling to really challenge and go up against it in any fundamental sense. Such people really don't want an *actual* revolution—so they also don't want to seriously engage questions of revolutionary strategy and possibilities and all that would be involved in an actual revolution, and they especially don't want to see lots of other people seriously engage such questions. Why? Because broad and widespread discussions and engagement of such questions would tend to undermine their own "credibility," such as it is, and completely cut the ground out from under that position they've carved out for themselves as *never anything more than* "critics" who are ultimately committed to remaining loyal to the existing system. And it's often these types who are some of the worst, when it comes to snark and gutter attacks, especially aimed at someone like Bob Avakian who has dedicated himself to an actual revolution and is actively working for and leading people toward that.

Serving Self, or Serving the Cause of Emancipating Humanity?

Question: Yes, and one form that this nasty snark takes is people who say, "Oh, BA is just doing this, or the Party just exists, because of BA's ego, to serve BA's ego." How would you address that?

AS: Well, to be honest, I would just start by saying that's so completely ridiculous that it's barely worth answering, barely worth dignifying with an answer. It's absurd, OK? Just think about it for a second. Somebody spends decades of their life dedicated to working on the most complicated problem you could imagine, which is figuring out how to make a revolution in a country like the United States, and leading others to be part of the process, working on involving broader and broader numbers of people in that process. Plus doing all this under the constant pressure of not only the threat of what can be brought down on a revolutionary leader by the authorities running the society, but also in the face of a lot of societal opprobrium—the tear-down attacks and facile dismissals without engagement, and all this constant snark and slander. **Why on earth would anyone subject themselves to all that just to satisfy their own ego?** I'm quite confident that BA as an individual could have had a much more comfortable life, a much easier life, if he'd just gone about his business as an individual looking out for "self," doing various other things, doing just about **anything** other than attempting to lead a revolution in the United States of America [laughs], and contributing to the international revolutionary movement. I have no doubt that, if he were just interested in having a comfortable life, or feathering his nest...there are a lot of other things he could have taken up or worked on...the idea that he would somehow be stroking his ego, and that the Party would exist to serve that vanity project, is completely absurd! And it's also frankly insulting, not only to BA but also to all the other deeply committed and self-sacrificing people who work with him and who dedicate their own lives to working for the revolution.

Look, it's once again like the question, the accusation, of religious cult. The fact is, there would be no great significance to BA as an individual if you divorced him from the question of "problem and solution," from the need for a revolution, the possibility of a revolution, the correct approach to take to making a revolution, and to building up a new society. If you separate him off from these types of questions, well then he's just an individual who would primarily matter, like most people, to his friends and family. There's no great significance to his role otherwise. **The reason he's significant** is precisely **because** he is a very advanced theoretician and practical guide for this revolutionary process—a guide who has actually brought forth a framework that, in some important ways, is **a new framework** for how to actually move beyond this stage of capitalism-imperialism and how to open up a new stage of communist revolution that would benefit the vast majority of humanity. **He's proposing a new model**, informed by a much more scientific method and approach than at any time in the past, and he's working extremely hard to get more and more people to check it out, to investigate it, to learn about it, to discuss it, and to hopefully "get on board" and act accordingly.

And the reason the Party, and supporters of BA who are not in the Party, promote him, is definitely not to promote some kind of religious cult. **This is not a religious cult, it's the furthest thing from it.** Again, to suggest that it is a religious cult, or any kind of cult, is disrespectful not only to BA as an individual who has worked so hard on all these problems for so long, and who has never sold out, as Cornel West often points out; it's not only disrespectful to BA, it's also disrespectful to all of the followers of BA, to all of the other people who are working hard in their own ways to try to be part of the revolutionary process and who find great value, great inspiration, and great insights in what BA has brought forward, and who are striving to actually get more steeped in the methods he has been modeling and developing in order to take a more scientific approach to the transformation of society, to the analyses of the problems, to the analyses of how you would build a society on a much better basis, and so on. So, it's disrespectful to BA's followers as well, who are the furthest thing from "blind faith" followers...who are people who care a lot in their own ways and who are also working very hard in their own ways, but who have the sense to recognize when someone has the qualifications to be a team leader and who appreciate that, derive a lot from that, without being slavish in any way, shape or form.

I mean, I consider myself an ardent follower of BA and a good representative of that, and I'll be damned if I'll be characterized as some kind of a "cult follower," or anything like that. I'm sorry, that's absurd as well.

Look, it's OK if you have **honest questions** about why people are working so hard to promote BA. For instance, you might honestly want to understand why people are wearing T-shirts with BA's image on them. Why are there graphics, posters, palmcards and websites promoting BA? Why did an experienced artist go out of his way to contribute those beautiful paintings with the graphic representation of BA? What is motivating some people even in other countries to want to wear BA T-shirts and to draft additional beautiful designs? Why do we advertize and promote BA's books and essays and films? Why do we organize meetings, programs and symposia around BA's new synthesis, or make a big deal about things like his recent Dialogue with Cornel West at Riverside Church? Why is there an ongoing massive fundraising campaign to make it possible to spread "BA Everywhere," to all corners of society? Why is all this activity taking place that is so focused on this individual? Well it's not difficult to answer such questions. It's basically because he's a very unique, very advanced, very developed revolutionary thinker and strategist. As I mentioned earlier, at this particular point in time there's really nobody working quite on that level, in the U.S. or anywhere else. He's definitely the leader of the field, objectively. If this were in a field of the natural sciences, and you had a chance to read or go see or listen to someone who was being described as the most advanced representative of that field of science at this time, you would have no problem understanding why that would have value, and why people would be right to spread the word about this individual and this individual's works, right? It's so funny, you know? People are always concocting and promoting actual "cults" of individuals in all sorts of spheres: What about the Obama cult, or the cult around various Hollywood celebrities, or popular sports figures, or whatever? Or maybe these days we should be talking about the widespread "cult" of the cell phone "selfie," what do you think? [laughs] Look, there are all sorts of stupid "cults" that get built up for marketing purposes, or to promote all sorts of superficial things, things that have nothing to do with trying to serve and benefit humanity. OK? So, if you're just trying to *honestly* understand why we promote BA as an individual, that's fine, *go ahead and ask your questions*. But those snarky people who claim there is a kind of "religious cult" being built up around BA, or that BA is just stroking his ego or something, don't have a legitimate leg to stand on. With all their obsessive snark and slander, one might even argue that they are falling into being typical anti-communist "cult followers" of a certain type themselves! [laughs]

But, more seriously, the main point is this: Why are there all these BA materials? Why does the Party work really hard to promote BA? And why do supporters of BA work so hard to promote BA, to build up the BA Everywhere campaign? Why do they do that? Simply because of the desire to spread BA's message as broadly as possible throughout society, to as many people and as many different kinds of people as possible. And then we'll see where this will all end up. If it turns out BA's completely off his rocker and not actually applying correct scientific methods to the analysis of the problem and the solution, then this will not go anywhere. OK? But, the fact of the matter, I'm firmly convinced, is that there is genuine scientific rigor to BA's whole method and approach, and there are implications and conclusions that are derived from that, which people really need to know about. The message needs to be spread far and wide. And people need to know that a leader such as this exists. They need to **know him**. They need to know the characteristics of his leadership. If it were just a question of spreading his name and image, with nothing much to back it up, there would be no point. But the reason to keep using **all** these materials and keep promoting his name, his image, and all his many books and talks and films, and so on, is **to get people to actually grapple with the underlying questions**. Because a revolution is not made by an individual, no matter how advanced that individual is. And people need to come into this process, and learn about it, and contribute their own insights. I'm sure BA would be the first to tell you that he didn't develop the new synthesis of communism in a vacuum, and it didn't just spring forth from his brain one fine day. [laughs]

It was developed on the basis of close and critical analyses and evaluations of the whole past experience of the communist project, informed by a close examination of the contradictions of the current world and current society, anchored in consistently sound scientific principles and methods, and further developed in relation to the ongoing motion and development of actual reality. And, in all this, it is clear that BA has also been more than willing to draw on, and incorporate, important insights and contributions brought forth by other revolutionaries and communists (past and present) as well as by all sorts of other people from every corner of society. Again, his whole method and approach is just **really good science**—it is the furthest thing from religion or any attempt to build up a religious cult!

Question: I also have to say this quickly: People have all kinds of T-shirts with all kinds of people on it, but when it's a revolutionary communist leader, all of a sudden people have a problem with it. It does really get to, and intersect with, anti-communism more generally. And the other thing I would say, kind of related to that, is I definitely agree that if somebody were looking to "stroke their ego," there would be a lot of easier ways to do that besides leading a communist revolution.

AS: And in a country like the U.S.!

Question: Not exactly the easiest way to get a lot of widespread praise in the short run.

AS: [Laughs] Well **that's** for sure! That's definitely for sure. But I think you're putting your finger on it. People are reacting this way—not all people, obviously, but some of these people who make these kinds of slanderous comments—**because they're out of their comfort zone**. There's something that disturbs them. So let's ask this: **Why** are they so disturbed? They're disturbed because here's somebody who is serious about revolution, and who actually has some real substance. It's not just a bunch of empty yak yak. There's deep analysis and deep synthesis on the nature of the problem, the solution, the direction of society. In some ways, some of these people are being left behind. Or, frankly, maybe the better way to put it is that they're having their raggedy asses exposed, in the sense that maybe they've been grumbling about some of the problems of society for a long time, but have never had much to offer people in terms of actual **solutions**. And BA does have a lot to offer, there's a lot of substance there. So, when BA is promoted and when he gets to be known better by more and more people, and his message is spread far and wide, then these other people who've maybe been full of bluster and been sort of throwing their weight around like maybe they knew some things...who've maybe been making something of a profession out of being just critics of certain things in society... all of a sudden, they're being called up short and maybe losing some credibility, because they don't actually have anything very substantial...they really don't have much to offer.

That's pretty common in any society, you know? You can always find a lot of people willing to be critics, to complain and grumble about various things and maybe even about the system itself; and some of these critics will sometimes even do some good exposure of some of the crimes of the system. But in the end, **what are they proposing**? What are they saying we should **do** about any of this? Once you've taken note of, and talked a bit, about the problems, then what? Where do you go? And usually, at that point, it gets to be kinda like, blah, blah, blah. **A lot of nothing.** All of a sudden, there's not much there, you know? I think we've all had the experience of listening to some people who can sometimes do some pretty good exposure—maybe even describing some of the abuses of this system in some pretty accurate ways—but then, pretty quickly, it all sort of dribbles off at the end. [laughs] There's nothing there! I find myself saying that all the time, after listening to some of these kinds of people—I'll say, "Well, this is unfortunate, because there's really nothing there! Sure they have some critiques, but they don't really have anything of substance to offer in terms of solutions or ways to fundamentally change society." But when you check out BA, it's a whole different thing. He's got a lot to say about the problems of society, but he's *also* got a lot to offer in terms of *solutions*—thanks to the work he's been doing over a number of decades. So there's a lot of substance there. And those of us who recognize this, and who are very much inspired by his consistently scientific method and approach, by his willingness to always look for the truth of things, no matter how unexpected or uncomfortable it might turn out to be, by his willingness to follow through on that, and to always proceed back from the need to emancipate humanity, and not just to advance the interests of one or another subsection of society...those of us who are inspired by the whole approach of consistently doing **that**, want more and more people to know about BA and to grapple with some of the fundamental concepts and methods he is putting forward. Therefore we want to spread knowledge about BA and his work far and wide throughout this society and, frankly, all around the world. So those of us who consider ourselves followers of that are not in any way defensive about that, and we have no problem explaining why this is so important and valuable. And, again, thank you to BA for making it possible for us to do that!

But another thing I'd like to say is that, given everything that's been going on lately, I suspect things are going to get better soon. Maybe I'm wrong. But I think that, since Ferguson, there's been a renewal of really good resistance and upsurge in terms of people coming together and uniting to say, "No more—we aren't going to take this any more!" Protests have spread, and have actually drawn in many of these middle class people who have joined in with the most oppressed and the most directly affected by the brutality and murders by the police. Large numbers of people have been marching through the streets of cities all around the country. And even in many cities around the world, people have marched and protested to support that. So this is an example of people raising their sights and starting to think in a more generous manner about the needs of the people who are oppressed and exploited... it's bringing out many people's "better side."

But right now there's still an incredible amount of passionate hatred towards Bob Avakian on the part of some middle strata people who, with very few exceptions, have never really even seriously read or in other ways engaged his work...his theories, his analyses, his strategy, his conception of the future society, his conception of how we could actually win, and what it would mean to win in a way that would mean something good for the majority of people, that would enable people to actually create a society that we would want to live in...that *most* people would want to live in. There are all these deep analyses and deep strategic thinking, but these people generally don't even bother to really check out and study BA's works. They're just dismissing him, and attacking him. And, frankly, I think a lot of it also has to do with fears they have of people at the bottom. That's one aspect. It's pretty obvious that Bob Avakian understands the people at the bottom, and he speaks directly and truthfully and honestly to people at the bottom. And people recognize his integrity. They know they're not being bullshitted. They can see that he "gets" their lives, that he gets their needs, that he gets their potential, that he gets their abilities, that he gets their value. He is speaking directly to them. And he is saying, "I know the way out! Listen, and become part of this, because this is, more than anything, for you and by you. And I have a plan, I have a vision. And what's missing is you."

He's speaking to all this with both heart and science. And people are responding to it. **People at the bottom are increasingly getting him.**

Look, Bob Avakian has written many complex books and articles that **do** require some serious study. He's also written many popular, more easily accessible things, and he's given a number of talks that people can easily understand and connect with. And what's interesting is that the people who often seem to **most quickly** "get" him—who most quickly "connect" with him, and with what he's about—are people from among the most basic sections, from among the most oppressed in society. In other words, from among the people who *most* need things to really, really, change. It's a little different among some of the most highly educated and "intellectually sophisticated" people—the ones who have been lucky enough to get fancy educations and have been trained in how to read complicated, abstract material: They're not always as "quick" to understand and connect with BA as people from the more basic strata. **Some** of those more educated people **are** "quick," and do "get" him and what he's about pretty quickly...some of them **do** get very enthusiastic when they actually seriously explore and engage his writings and other works. And of course it's generally easier for such people to read and study the more complex materials because of their training and background. But some others from among those educated people, well...how can I put this? I would say that, despite all their education and other advantages, their hearts may not be quite in the right place, or they may be conflicted about how much things really need to change, and what kind of world we should all want to live in. So that's an interesting difference. Again, it's one of the historical characteristics of the petit bourgeois class in an overall sense to be divided and torn like this, and to be trying to put the brakes on...On the one hand, they're essential as part of a revolutionary process—revolution can't happen without many of those people getting on board. And, on the other hand, they're often trying to drag it off course, or put some obstacles in the way, or stop it altogether. So people have to be struggled with.

You know, in terms of the problem of the snark culture these days, I said before that I suspect, and I certainly hope, that this is going to start to change soon: If people are struggling and resisting and becoming more part of a revolutionary process in this period, I think there will be more people who will want to turn to all these haters and promoters of snark and say to them, "You're disgusting, you disgust me! You're boring and unimaginative, you're ill-informed, you have nothing to say, you have no answers to the problems that we're wrestling with, you have no answers to how to get out of this mess in any kind of a deep way. So cut it out!"

Question: And, as you've said, in a lot of cases, these people don't even pretend that they've engaged BA's work.

AS: No. And let me tell you a story about that, a true story. A really outrageous thing that happened after the Dialogue at Riverside Church. First of all, let me say again that I personally feel very fortunate to have been able to be at that Dialogue between Bob Avakian and Cornel West—it really was very rich, and I felt both speakers contributed a lot. I'm obviously a follower of BA and I drew a lot from his presentation in his part of the Dialogue, but I also learned a lot from Cornel, including his modeling of principle and integrity. I learned a lot basically about morality and conscience as embodied in a progressive religious person who maintains his differences and does not hesitate to bring them forward, but actually promotes a very moral and high-level conscience and social responsibility. And he said some very important and interesting things in the course of that as well. So while all that's going on, I'm thinking, "This is a tremendous experience, this is a wonderful atmosphere, there are all sorts of different kinds of people coming together to be part of this, there are 2,000 people here and they're going to spread the word about this and more people are going to be watching the film of this in days to come, and this is a wonderful thing. It's a great victory and advance!" So I was really happy about this. But then you hear that there were a small number of people there who were complaining. And what exactly were they complaining about? "Ehhhhh, Bob Avakian talked too long—and, oh yeah, this was disrespectful to Cornel West." Well, in my opinion it's really condescending and disrespectful to Cornel West to think that he needs to be defended against the big bad Bob Avakian speaking too long, or anything like that! Cornel West is perfectly capable of getting his message across, thank you very much, and he definitely succeeded in doing so at the Dialogue. Bob Avakian spoke for some time because he had a lot to say, and most people there wanted to hear it. This was a particularly rare occasion, and there were some people who came from very far away specifically to hear Bob Avakian, and they wanted to hear him talk as long as he was willing to talk. But a small number of people were just raising these types of petty complaints. Could **they** have articulated a genuinely different and comprehensive program, a truly radically different approach to social change? No. But they **can** complain, "Well, he spoke too long, you know, he hogged the time," or something like that. That's literally been the focus of their criticism, very low-level and petty. To me that's really pathetic, and all the more so in the context of these times, when there are so many truly big problems to address.

But I have heard of things that are even worse than that. I'm going to tell you that story now: I happen to know for a fact of **at least two cases** of people who are sort of academic professor types, who did **not** attend the Dialogue, but who nevertheless proceeded to rant and criticize it after the fact, going on and on about, "Oh Bob Avakian this, Bob Avakian that," in a very negative way. Two different people, in two completely different places. And when they were asked, "Well, given that you weren't there, did you even **listen** to the recording of the Dialogue, did you even watch the live stream of it, since it's available on the website?" And in **both** cases, they replied, "No, I didn't watch it, I didn't listen to it...**because I didn't need to**." Wow. Really? They're trashing it, spreading really negative criticisms all over the place about something they did not attend, did not listen to, and never bothered to watch? Again, wow.

Now, keep in mind, these are actually *professors*. Imagine if they had a student in their class who wrote a paper about something, critiquing a book or a film or something, and the student were asked, "Well, did you actually read the book, or watch the movie?" and the student responded, "No, because **I didn't need to**." Don't you think they'd be inclined to reject that paper and maybe even flunk that student out of their class? [laughs] And yet, somehow, it's become acceptable among some of these types to do this kind of petty, low-gutter, disgusting sort of attack and criticism—it's very shallow, very petty, and very low on substance—and then you ask them, "Well, did you at least listen to the audio, did you watch the live stream?" "No, I didn't need to." Well, how about Bob Avakian's many books and other works? Have you read **any** of them? Which ones? What do you think of the way he puts out the strategy for revolution? What do you think about his analysis of the key concentrations of social contradictions in the U.S. today, and of what he says about how revolutionaries should work on these to advance the revolutionary process? What do you think about his analysis of the need to proceed back from strategic objectives rather than just trying to develop the movement at any given moment in little bits and pieces? What do you think of his vision for how you might actually be able to go up against the armed force of the capitalist state when the conditions are ripe for revolution and actually have a chance of winning, because of the approach that you'd take when it came to that (an approach that is outlined in "On the Possibility of Revolution")? Have you read that? What do you think about it? What do you think of his vision for how to structure a new society, the institutions of law and the courts, people's rights, and elections, the role of elections? What do you think of his discussion about why there shouldn't be an official ideology in socialist society? What do you think about **any** of these issues? Do you have *anything* of substance to say about any of that? I mean, I'm just doing a random selection here. There are many, many such questions that BA addresses but that most of these people have never looked into or even bothered to think about. What do you think of his analysis of the problem and of the solution, and of the strategy and the guidance that he's giving to the practical movements on the basis of a theory that's been developed and been refined over decades, and which has taken the best of the past experiences and sorted them out and recast them to bring forward a new synthesis of revolution and communism? There's a lot of work there. What do you think about any of that? "Ehhhh, bah, bah, bah, bah." [both laugh]

These people don't have anything to say about these important matters, matters that could have big implications for the lives of millions and billions of people on this planet, but somehow they think it's OK to spend a tremendous amount of time grumbling among themselves and using the web to spread these kinds of hateful little petty, petty disses against this revolutionary leader, and the revolutionary movement. As I said before, ultimately, it's not just a disrespect to Bob Avakian. It's a disrespect to the people. It's a disrespect to the people at the bottom who desperately need revolution. It's a disrespect to the people whose children's blood has been running down the streets, and who can expect more of the same, day after day, year after year, until this system is overthrown.

Question: I think that's really right, because when they're making these personal attacks and petty slanders or dismissal without engagement, they're directing all this venom against this revolutionary leader who, as you were just saying, has a way out, has a way to emancipate not only the most brutally oppressed under this system, although very critically them, but all of humanity. They're spitting in the face of that way out, and of the possibility for a whole different society and a whole different world, including for those for whom the world really is horrific, full of horrific suffering. With a lot of these petty slanders and this snark and nastiness, and with the way that's related to the class outlook of the petite bourgeoisie, or the middle strata, it seems that, as reflected in the story that you were just telling, there's just a lot of arrogance: "Oh, I don't even need to check this out."

AS: Right.

Question: Well, why would you **not** need to check that out? So, you know everything about why the world is this way? When you say, I don't need to check it out, either you're saying, I know everything about why the world is the way it is and how it could be different, or you're saying, I don't care. Or both.

AS: Right. Right. Or both.

Question: So either way, it's arrogant. But then, beyond the arrogance, I think that merges with being very uncomfortable about what Bob Avakian and his work actually represents. In other words, to put it bluntly, being very uncomfortable with the idea of revolution, with communism, with the dictatorship of the proletariat—and actually with certitude, with saying, I know the way out, I have done this work, and I have developed a way out. So maybe you could talk about how those two things coming together: **the discomfort with revolution and communism, and the discomfort with certitude**.

AS: Well, that's a good way of putting it, because I think it gets back to this point that, for some of these middle strata, they've got one foot at least in still wanting to preserve this system and their position within it, even if part of them wants to get to something better. And they're constantly driven to explore very low means of trying to change things. For instance, they keep being drawn again and again and again into the *illusions of elections*. I mean, you would think—most of them are not idiots, and you would think that, after banging their heads against the wall over and over and over again, they would learn a lesson or two about how elections are no way to fundamentally change anything, or even to get rid of some of the abuses and injustices that they would actually like to see eliminated. You know, there were all these people getting swept up in the Obama presidency and thinking that it was really going to make a difference. And Bob Avakian and the RCP kept explaining, "No, look, you're just getting caught up in a trick here. It's not about the individual, it's about the individual playing a role as a **representative of this capitalist-imperialist system**. It's not going to be changed by having a different color president." But, at first, almost nobody wanted to listen to any of this. And of course, now that it's become more obvious that nothing is really being changed in any meaningful and positive way, it's not like people are saying, "Oh, well, I guess you were right in your analyses and maybe we should pay more attention. Let's go back and talk about it." No, now they just want to move on to the next question, the next illusion. They keep getting caught up in illusions, in a lot of reformist schemes.

A Need for a Big Societal Debate: Reform, or Revolution?

AS continues: There's a big mass societal debate that needs to go on among all strata on the question of reform or revolution: Which is the way forward? Reform means you tinker with the system, you try to fix it here or there. An example of reform, for instance, is that you would try to deal with police brutality and murder by things like having civilian review boards, and putting body cameras on policemen, and in other ways trying to fix things *within* the existing relations in society, *within the existing system.* And people do try these things. Civilian review boards have been around since at least the 1960s, you know. People keep falling into these traps of these reformist notions that, somehow, if you could just fix and tweak this a little bit here and there, you could get rid of these outrages and abuses. And what that comes from is a lack of profound scientific understanding of why these kinds of outrages are not just accidental or occasional, and why they're deeply rooted in the fabric of this system, in its very foundation, which has everything to do with the white supremacist origins of this particular society in the United States, how this country was founded on slavery, and everything that came from there that's never been surpassed. It's not just a question of backward racist ideas on the part of some white people. That is in the mix. But much more deeply, **there is an institutional fabric in how this capitalist-imperialist system is structured and how it works**, in such a way that it cannot resolve these deep, deep divisions and problems, that requires that certain sections of society be kept down and oppressed, in particular, Black people in this country, and other people of color as well.

That's a whole bigger discussion, there's a whole deep analysis of why that's true. This has been deeply gotten into by Bob Avakian and the RCP, and people should check that out. I'm not going to try to get into it more here. But in the context of what we're talking about right now, I'm saying that, if people are just looking for some ways of making a few token **reforms**, a few "tweaks" to the system, or looking for ways to maybe improve a few things in just one neighborhood or local area...it's not that all those kinds of projects and plans are really bad in themselves, but it's that they **won't lead to the fundamental change that's needed**. For instance, look at the environmental movement. What is happening with the environment is a global emergency that requires big-scale measures of restructuring the way the economy, and society overall, operates, to prevent the constant exploitation and degradation of the environments of the planet that's going to end up leading people to extinction, you know. I firmly believe that humanity is either going to find the ways to transform its forms of social organization in the direction of viable socialism and eventually moving towards planet-wide communism, or humanity's going to go extinct because of what it's doing to this planet. I can make scientific arguments about why I think that's really true. And time is getting short. So that's just one example: Why the environmental problem has to be tackled on a really big scale, by making really fundamental, radical change in the whole way society is organized, structured and run. Just a little more enlightenment and just a few tweaks and minor reforms of the existing system are simply not going to cut it.

But a lot of progressive-minded middle strata people...often they'll get into things like, "it might be better not to use plastic bags at the grocery stores," or "let's have green light bulbs," "let's recycle more," or "let's see if we can work on hybrid and electric cars to reduce pollution, and let's have more solar panels, for clean energy." There's actually a lot that can be learned from a lot of these initiatives, and many of those kinds of changes **are** things that you would actually want to implement in a new society. And I'm not saying that it's bad to be encouraging some of those small steps even today. But what I would like people to recognize more honestly is how puny, limited and tokenistic these changes are, especially relative to the actual scope and scale of the environmental crisis. **It's not even scratching the surface of the problem.** What is needed is much more profound, radical change. And I think a lot of the middle strata people are always looking for these "little ways of tinkering," trying to reform just a few things, in a way that seems more comfortable and manageable, rather than confronting the need for a total dismantling of the system and institutions that are necessarily **driven**, by their own **underlying laws of functioning**, to despoil and degrade the environment. **The system of capitalism-imperialism cannot stop doing this, it is structurally unable to stop doing this—that's what you have to confront.** People sincerely concerned about the global environment really should seriously study Bob Avakian's analyses, which make the case, and provide evidence, for why problems such as these are so deeply rooted in the very functional core of this capitalist system that they can't be dealt with just through a series of minor adjustments. What is required is a profound and radical overhaul of the whole way society is set up at its foundations, of the whole way it functions in a comprehensive sense. But to effect this radical restructuring it is necessary to have an actual revolution—so **<u>that</u>**, more than anything else, is what people genuinely concerned about the global environmental emergency should be working towards.

These arguments are backed up by a lot of sound and concrete scientific evidence. Nevertheless, a lot of these middle strata people are uncomfortable with the prospect of such radical change. In some cases, it's more that they just haven't yet encountered these analyses, they're unfamiliar with them, nobody's ever talked to them about this, they haven't yet explored the revcom.us website and *Revolution* newspaper or the works of Bob Avakian. But I'm sure many of them—especially among the younger people who are not so invested in reformist methods and approaches—will find their way to these resources, and will start seriously digging into all this themselves, and I think many will end up being willing to confront "the logic of the logic." In other words, when they seriously dig into the analyses, they will increasingly recognize that, "Yes, this does makes sense, this is what the evidence points to." And even though revolution is not an easy road, and there will necessarily be sacrifices, it would all be worth it to have a genuine possibility of making a much better world, of constructing much better societies, on a new basis and foundation that could very quickly address the major problems of capitalist society, and which would greatly benefit the vast majority of people. The irony is that all those middle class people who constantly complain about the way things are today but who shy away from radical change and revolution...many of them, most of them in fact, would, I am quite sure, end up very much benefitting from, and appreciating, life in a new socialist society, especially a socialist society of the type envisioned by Bob Avakian's new synthesis. Once again, on the foundation of that solid core, but with lots of elasticity based on the solid core, there would be air to breathe for these people in such a society. They would not be pushed to the side or crushed or stifled, as long as they were not trying to actually destroy the new society, and they would find that they could themselves help institute a lot of the progressive social changes that they get so frustrated at not being able to implement under conditions of the current system. So they should look forward to it, and help work towards it.

But again, right now, especially among the middle class, a lot of these people are more inclined to stick with what they're more familiar with—the known rather than the unknown. They haven't dug into any of this, really. They haven't checked it out. They haven't discussed and debated it. Many seem more content with puttering around with little reformist schemes, making little minor criticisms, and just basically complaining about the way things are, but without really doing anything that's substantial to get beyond this. And the crime is that, meanwhile, while they do that, while they cultivate and promote their illusions, and when they try to tear down a revolutionary leader like Bob Avakian and try to prevent him from getting his message out broadly to the people—while they're doing all that, the world continues as it is, with the unrelenting grinding down of the masses of people here and around the world. The blood and the bones—this is real, it is ongoing, and it will continue to go on, on a daily basis, as long as this system is allowed to persist.

Different Positions in Society, Different Views of Revolution and Revolutionary Leadership

Question: I think a lot of what it comes down to is how much of a stake in this capitalist-imperialist system do people feel they have, because, as a kind of contrast, I saw a comment from somebody on the website revcom.us—leading into the Dialogue between Bob Avakian and Cornel West, there was somebody who said: I'm not a communist, but I'm not an **anti**-communist. I read that and I thought, there needs to be a lot more of that spirit among people who don't consider themselves communists, that spirit of openness. In other words, on the one hand, as the people who are brought together as part of the Dialogue and as the exchange between the two speakers in the Dialogue shows, the point is not that, unless somebody is a revolutionary and a communist, they won't appreciate BA; but, on the other hand, if somebody feels that they really have a stake in this system, and they're very committed to the idea of trying to "make things work" for them within the system, then it would not be surprising if they had a lot of antagonism towards BA. Because, as you're saying, very quickly when people engage BA it becomes clear that he's going very vigorously and very consistently and uncompromisingly against this system and insisting, as you were saying earlier, that no fundamental change is gonna happen and this needless suffering is not gonna end until this system is swept away through revolution. Well, if you're threatened by the idea of this system being swept away through revolution, because you're still trying to carve out a little bit of space for yourself within this system, of course you're not gonna like what BA stands for and represents.

AS: Yes, I think there's that, but there's also a lot of ignorance and fear and misconceptions about what **is** a revolution, what **is** a communist, what **is** leadership, what does all that involve. For instance, a lot of these middle strata people seem to think that leadership is inherently a bad thing. They say, "I don't want to be led!" Well, guess what, **you're already <u>being</u> led**, and on a daily basis, by the rulers of this system. So get over it, OK? The questions you should be asking are: **What kind** of leadership is being provided? **to accomplish what? for whose benefit?** Those are legitimate questions. And then I also have to ask such people: If you aspire even to some degree of social change or even radical social change, how serious are you willing to be? Because a revolution is a serious thing. It's a complex thing. Radical change, radical transformation of society is a complex and serious matter. And, if you want to seriously undertake that kind of radical transformation, you need to realize that you will have the responsibility for the lives of millions of people in your hands. So, by all means, take a serious look at it. Ask those questions about what kind of leadership is being provided, working towards what objectives, and to whose benefit. It's like that statement on the revcom.us website: **What is needed is an actual revolution, and if you're serious about an actual revolution, you should seriously get into BA.**

So, make a start, if you're serious. You don't have to be a committed revolutionary communist, you don't have to agree on everything, you can just start checking it out. It's worth pointing out, for instance, that some of the people who are helping to broadly promote BA as part of the BA Everywhere campaign don't consider themselves at this time to be communists, or even revolutionaries. What they have in common is that they know how to recognize a leader who can be a reference point, and they see the value of spreading the genuinely substantial works and the analyses, in order to get more and more people in all corners of society grappling with all this, discussing and debating the big social problems of today, their underlying causes, and what BA is putting forward as a solution. Through this whole process some people will no doubt decide, but at least on a more substantial basis, that they don't agree and don't want to be part of this. But many people will say, "Well, this isn't what I expected it to be, I didn't expect BA to be like this, but so far I like what I'm seeing...I didn't expect to be interested in revolution and communism...I was taught differently in school, or in church, and so on...so I thought it was a really bad thing. But now that I'm learning more about what it's actually about, and I'm learning about BA's plan, and his approach to things, the way he thinks...and getting a feel for what the new society could be like...I am finding that I am appreciating more and more of what I'm seeing and hearing."

So all this is a process, and it's also a matter of people finding their place within it.

For a lot of the more basic people in society, the problem is they think they're too messed up, or they're too uneducated, or whatever, to be part of this. But that's completely wrong. They are going to actually play a very major role in the revolution, a key and pivotal role, a core role at the foundation of the revolutionary process. Why? Because they're not so inclined to try to keep a stake in this system. Why would they be? Most of them, objectively, don't have anything to preserve in this system, in this way of life. So, in a way, that frees them from a lot of those entrapments that the middle class people fall into. The basic people, including those at the very bottom of society, have a great deal to contribute to the revolutionary process—their ideas, their abilities will help to develop the revolutionary process and the revolution itself, and to build the new society.

But this gets me back again to the fear and misconceptions that many people have, particularly in the middle strata. We've talked about how, if you have a big stake in preserving something in this existing system, then maybe you're not going to like BA. [laughs] But there's also the fear. It's not only that there are prejudices, misconceptions and fears about what communism and communist leaders are all about; I think many of these middle class people are really afraid of people at the bottom. And there's also some fear of what it might mean for the two to increasingly connect—the revolutionary communists and the people at the bottom. They can see that BA directly speaks to and connects with the people at the bottom, the most oppressed, the people in the inner cities, the kind of people that you see coming forward in a place like Ferguson after the murder of Mike Brown. And some of those middle class people are frightened by this.

Question: BA says, **these are our youth**.

AS: Yes, these are our youth, and he means it. And he's calling them forward. He's giving them a way out, and he's giving them a way to become part of the process and to actually develop not just as cogs in the wheel, not just as participants in the revolution... with what BA's brought forward, there's a way for people to come forward from those kinds of backgrounds and develop not just as **contributors** to the revolution, but as **leaders**—to become leaders themselves, leaders of the people and leaders of the revolutionary process. Again, that was part of the point of the example of Wayne Webb, whom people also knew as Clyde Young. He could have spent his whole life in prison. Like many Black people, many Black men in particular, he was on track to just be thrown into the dungeons of the system and basically spend his life going back and forth between the mean streets and the prisons. That could have been his life, like it is for so many people. Revolution and communism and the leadership of BA actually **drew him out of that**, and gave him the **foundation** and the **basis** to develop, not just as a revolutionary but as **a revolutionary leader in his own right**, who could influence in positive ways other people and help to build the revolutionary process. That sort of thing happening on a large scale would be a nightmare for the ruling class. The bringing together of genuine revolutionary communists and the leadership of someone like Bob Avakian with the masses of oppressed people, and in particular oppressed Black people in this country—that is in some ways the worst nightmare the ruling class could imagine.

It's one thing for ruling class figures to have nightmares about that connection, and that development, but it's pretty disgusting when some of the middle strata people who imagine themselves to be liberal or progressive are themselves feeling that kind of fear or reluctance, and are getting in the way of that developing, and are trying to stop it. And one of the things about a class outlook is the phenomenon of people trying to remake the world in their own image, in line with their own interests, which is not necessarily the same thing as the interests of all humanity. See, that's a big difference between Bob Avakian and a lot of these middle strata people. He doesn't try to say, "Well, this is the way **I** would like things to be, this is what **I** need, what **I** would appreciate, and I want everybody else to take **that** up." Instead, he proceeds from a materialist, scientific analysis of what are **the objective needs of the vast majority of humanity**, not just in this country but all over the world—he proceeds back **from that**. How do we move to meet that? How do we move towards emancipating all of humanity, not just trying to do what I would like for myself? But a lot of the middle class people, they more imagine that whatever **they** would like to set up is what would be best for the world, or best for society—and that's not necessarily the case. And all this certainly needs to be subject to scrutiny and discussion. But they're removing themselves from the process when they don't engage.

Question: Well, I think it would be worthwhile going further at this point of class outlooks, and the class outlook of the middle strata in particular. And just to highlight a distinction you drew before, I know this is something people often get confused about:

As you were saying, the point is not that everybody individually who comes from a middle class background is going to think that way, or think the same way; but, to pick up on a point you were making earlier, the middle class is caught between the proletariat and the bourgeoisie, between the most oppressed and exploited class and the capitalist class, and people in this middle class don't want either class ruling over them—that's the class outlook of the middle class. And what goes along with that is the idea: Let's just make everyone equal. In terms of what you were saying about remaking the world in the image of the middle class, that outlook is "let's just make everyone equal." I think that comes into play in a lot of this. Along these lines, here are some types of things people with this middle class outlook often say. I'll put out a few of them, and then maybe you could respond. "Why do you make such a big deal out of this one person, BA? Shouldn't everyone be equal? Shouldn't the goal be democracy? Aren't we all leaders, shouldn't everyone's ideas be given equal consideration?" These are the kinds of things that often get said by people coming from this outlook. How would you respond to that?

AS: Why do we make such a big deal about this one person, BA? Well, it's not a very complicated answer. You have a person who's the most advanced theoretician of radical change, of revolutionary transformation of society, alive on this planet today. It's really that simple. He's not a god, he's not some kind of mystical cult leader, or anything like that. He just happens to be the most consistently scientific and advanced person in terms of learning the lessons of the past, both positive and negative, and drawing from many broad sources, to forge a whole new synthesis, and provide leadership, theoretically as well as practically, to get us out of the situation we're in, in terms of the major problems that plague society and the world today. He's got the analysis of the problems and the analysis of the solution, and he has the ability to lead people in applying the science to concretely transforming things in the direction of revolution, the kind of revolution that actually serves the emancipation of humanity and is not just promoting one group of people over another, or creating a new class of oppressors to oppress people.

And this question basically has a lot to do with, do you think this world, as it is, is a horror? Do you think this system creates an unbelievable amount of unnecessary suffering for people, both in this society and around the world? If you do, what are you going to do about it? And what do you think is involved with that? Do you think it's just a simple thing of giving a thumbs up or a thumbs down, or doing a "mic check" or something? [laughs]

I mean, it's ridiculous. It's ridiculous, it's frankly childish, when people don't understand the need for leadership. This question, why do you make such a big deal about this one person, BA, actually has another question hidden inside it: Why do we need to have any leaders **at all**? It has to do with the notion that a lot of these people have, that they don't want **anybody** leading them. That is a kind of a childish protest, that corresponds to the middle class outlook of people who are filled with illusions, imagining that somehow they are some kind of free agents in society who manage to escape being led by anyone. That's so absurd.

What do you think a society *is*? A society, any kind of society, involves leadership. In a capitalist-imperialist society—of course, there's leadership! You're being led on a daily basis. How do you think things happen, like the production of things, in a society like this one, or on a global plane? How does agriculture, and industry, and transportation, and health care, education, all the various institutions...how do you think they get built up, and on what basis, with what values, with what objectives? Do you think they just kind of emerge spontaneously, like mushrooms after a rain, and that they require no leadership? That people just bop around day to day, running things, in a random and disorganized manner? Obviously not. So, of course, they're led. They're led with certain objectives. They're led with certain principles. They're led with certain values. The basic institutions of society under capitalism-imperialism are led primarily on the basis of generating and expanding profits, and out-competing rival capitalists. The way this kind of society is organized and led has nothing to do with actually being geared to meeting the needs of the people. That's not the basic objective of capitalists and it's certainly not their priority. From their standpoint, if they sometimes meet some of the needs of the people in the course of doing what they've got to do, then fine—but again, that's not what they're primarily geared to doing.

Does anybody really think that any major institutions, in any kind of society, can exist without some kind of leadership? And what do you mean if you say, "I don't wanna be led, I don't wanna be told what to do?" What are you thinking? That you're just going to get together with five or six friends and you're going to make all the major decisions of life? Think about what you are actually going to encounter in society: You're not going to be isolated on some kind of mountaintop. What about when you relate to **other** people and those other people have **different** ideas of what **they** want, or what's needed? Who's going to sort all that out and figure out what's actually the most correct thing to go for or the most important things to prioritize? On what **basis**, and with what kinds of **methods** and **orientation**, are such things going to be decided? You know, even in socialist society there will still be contradictions among the people. One example that has been given is this: What if **you** think we should build a park, and **other people** think we should build a health clinic? Maybe both ideas are good and both things are needed, but we just don't have the resources to build both at a particular time. Who's going to sort that out? Who's going to figure out what to prioritize? Who's going to make the decisions about what contributes to the direction that society needs to go in overall? Now think about how these kinds of questions will be multiplied many, many times over, and on a daily basis, if you're talking about the society as a whole. So how can all this be figured out and handled in the best possible ways? It will take leadership.

Earlier, I mentioned the "4 Alls." This is something from Marx that was popularized by the Chinese revolutionaries who followed the leadership of Mao. It is a concentrated way of describing the strategic goal of communism. It refers to the fact that, in order to get to communism, we need to move towards a world where we can get rid of all the class divisions among people, on a worldwide level; and, in order to do that, we have to move beyond all the economic-production relations that are the underpinning of those class differences. Because those class differences don't appear like mushrooms after a rain either. They have something underneath them, they come out of particular economic-production relations that foster those class divisions. And the idea of moving towards communism is, if you want to get rid of **class divisions** among the people, you've got to get beyond the existing **economic-production relations that are the material underpinning of those divisions**. And you have to get beyond the **social relations** that flow out of, or correspond to, those underlying production relations, the oppressive social relations like the contradictions between men and women, and between different nationalities, and between different parts of the world. And the fourth of the "4 Alls" is revolutionizing **all of the old ideas and the ways of thinking** that go along with relations of oppression and exploitation.

All this is a completely, a radically and fundamentally, different vision of how the world could and should be. **It's a completely different framework.** And when we talk about how **"BA is the architect of a completely different framework"**—of revolution, of the revolutionary process, and of the new society to bring into being—that's exactly what he is. You can like it or not like it, agree or don't agree, but **objectively** that's what he actually is. And he's been developing this framework very systematically, on the basis of scientific methods. And **that** is why we make such a big deal about this one person, BA. There is no one else in the world today who is on the same level in terms of developing the science of revolution and its application to the struggle to transform this society and the world on a radical, a truly radical basis, that deals with fundamental problems. Nobody's taken it as far, and on such a consistently scientific basis, and has as worked out a sense of not only **why** it needs to be done, but **how** to do it, and **what** to bring into being to replace this system. That's why we make such a big deal about him.

The way some people make a principle of opposing any and all leadership, insisting that they are "against the very idea of leadership" is truly absurd, especially when anyone who thinks about it for two minutes should be able to recognize that all the major components, of any large and complex society, obviously have to be led in order to be functional.

And then there's this other idea that some people put out: "Isn't everyone equal?" Or shouldn't our goal be to "make everyone equal"? Why do people say such stupid stuff?! [laughs] Look, it's one thing to say that all human beings are "equal," in the sense that every human being is a full human being and should be recognized as such. There's no such thing as an **"illegal"** human being, there's no such thing as a human being that's only **"one-half, or three-fifths,"** of a human being, there's no such thing as some kind of inherently **"inferior"** human being. **All** human beings are **full** human beings. That's one thing. But when somebody poses the question, shouldn't everyone be equal, what they're really asking is shouldn't everyone be able to throw their weight around to the very same degree, shouldn't everybody be able to have the very same influence on things? Well, that's not reality. I don't know what kind of dream world you live in, but the reality is that different people in human societies have different degrees of influence, **for good or bad reasons**. You know, there are some bad reasons why some people have disproportionate weight and influence. For instance, the people who run the government, who run this society, who run the police and the military, you're not equal to them. OK? [laughs] The bosses where you work, who have the ability to throw you out on the sidewalk, you're not equal to them either. Not because you're a less valuable human being, but because you're objectively not equal to them in terms of the social position you occupy and the influence you are able to wield. So these are examples where you can see that everybody's not "equal," since some people clearly wield disproportionate weight and influence of a negative nature.

The other side of this is that there are also people who wield disproportionate weight and influence of a positive nature, including in ways that can contribute positively to society, that can "serve the people" in various ways. Think of people who are "tops" in their fields, like a "top" doctor or lawyer or a "top" auto mechanic or a "top" athlete or musician. I don't think of them somehow being "better" human beings than me, but I have no problem acknowledging that I don't have their skills and experience in those fields and therefore that we are not all "equal" in that sense and therefore I wouldn't expect to be wielding the same degree of authority or influence as those "top" experts in an operating room, on a basketball court, or on a concert stage, just to use those examples. But I'm not worried about that. I don't feel threatened by that. We don't need to be "equal" in every dimension of life. And the reality is we're not all "equal" in terms of experience, skills and abilities. And in relation to positive things, it's OK, it's more than OK, if some people can wield more weight and influence. Which gets me back to BA. It's not only OK, it's more than fine, if BA is able to wield disproportionate weight and influence inside the Party he leads, in the larger movement for revolution, in the broader society at large. If he has the experience, skills and abilities that put him at the "top of the field" with regard to the analysis of the biggest social problems of this era and what to do about it, if he is objectively at the "top of the field" with regard to the development of the science of revolution and communism, then I, for one, want him to be able to wield as much disproportionate influence as possible! [laughs]

Nobody else has done this kind of work. Look at the work he's done. Be scientific and look at the evidence. Show me anybody else who's done this type of work, at this level, with this degree of depth and substance and innovation, anywhere in the world, anywhere in this society, or even in his own party. As I said earlier, he is clearly miles ahead of even the best of the rest. That's just a fact right now. So here, too, we're certainly not all "equal," we don't all have the same expertise and abilities. But why pretend otherwise, or act as if that's some kind of big problem? It's not! In particular, to everyone who sees the need to radically transform society and work for the emancipation of all humanity, I would simply say the following: 1) we **all** have the ability to make important contributions to the overall collective process of revolution, so let's get on with doing that; 2) we should all be trying to learn as much as we can, in particular about BA's overall method and approach, to work on "catching up" with BA as best we can, so that we can keep improving how well we function as a "team"; and 3) if we're serious about wanting to make a better world, we should recognize how lucky and privileged we are to have the opportunity to learn from such a "top expert," to have the benefit of such advanced leadership. And we should take full advantage of this.

So again, the fact that there are differences in people's abilities is just a matter of reality. And, I'll tell you, this idea of not wanting to recognize that some people have more influence than others, and that in some cases it is a very good thing that some people can have more influence, is a classic middle class kind of complaint. The people who are more at the bottom of society don't generally have this problem. They know damn well that people have different abilities, different experiences, different levels of expertise. They already know that some have had the privilege to have higher levels of education and advantages in life, and that some have had very big disadvantages in life. And they know that all this makes a difference to what people can **do**. Again, "equality" is one thing, if you're talking about how lives matter, and how all people are full human beings. But people occupy different material positions in society, concentrate different levels of expertise, wield different degrees of influence.

At the same time, when we talk about some of those misconceptions and wrong views commonly found among middle class people, let's not fall into being narrow or mechanical about this. There is not just one single outlook that "automatically" and inevitably comes along with your position in society. I don't want to encourage the development of "revenge thought" towards middle class people. Just because a lot of middle class people frankly have their heads up their asses these days doesn't mean that people at the bottom should want to chop their heads off! [laughs] Because many of these people can turn out to be very generous-minded, and very inspired by the developing movements of resistance, and by the growth of the revolutionary movement and its communist leadership, and many will contribute in some very good and very significant ways to the transformation of society in the interests of the majority of people. So, let's make sure we deal with things scientifically, by looking at how different people actually are, how they actually think and act, and not go around imposing generalities on people.

But, without falling into mechanical thinking, it is a fact that people do occupy different material positions in society, and that does tend to affect, it does tend to strongly influence, the outlook and the views and analyses of people. So, in a society like the United States, with its large middle class, you end up having people saying things like, "Well, I don't wanna have leadership." Again, as I was saying earlier, this sounds childish and stupid, frankly, if you think for two minutes about what's actually involved in running any part of even the existing society. And then, what about running and directing, providing guidance to, an actual **revolutionary process**, with all its complexity, with all its diversity, a process which ultimately needs to involve thousands and millions of people...many different kinds of people who come from many different backgrounds, who have many different ideas about how to go forward. If you're leading a revolution, you're gonna be trying to shepherd **all that**, and in a certain **direction**. You're gonna be trying to direct it toward **strategic goals**, that will benefit the majority of humanity. But, of course, not everybody involved is going to understand all of this in exactly the same way and all at the same time. In fact, they won't. This gets back to the example we were talking about earlier, about how, when you get to the point where there can actually be a revolution in a country like the United States, the majority of people involved in making the revolution will still hold on to religious ideas. They won't have all broken with religion. But, at the same time, they'll still be part of the revolutionary process. This contradiction is something Bob Avakian spoke to in the Dialogue with Cornel West. It's a good example, but just one example, of the complicated contradictoriness and diversity of views, and so on, that will come together in the revolutionary process, and that you can already see coming together in beginning ways.

But, if you want any process to go in a certain direction, and to accomplish certain goals, then you definitely need leadership. If you are trying to solve a problem in the natural sciences...let's say that you want to investigate the surface of Mars, and you get a bunch of scientists together...if they can't agree on the strategic objectives of their project, if they can't agree on the methods and approach, if they're all going in different directions with all this, you're just going to have a mess! [laughs] The different people involved might all contribute some important ideas and insights, but if there's no way of coordinating it, if there's no way of channeling it, or directing it, towards a specific goal—in short, if there's no leadership—you're not going to get very far. You need leadership to enable you to sort out what's right and what's wrong, and to actually make sure that, even if there are inevitably going to be some false starts and some dead ends, most of the efforts are at least attempting to go in a certain direction, towards a certain goal. So, of course, you need leadership. And who's providing that kind of leadership for the revolutionary transformation of society? Show me anyone else who has done the kind of extensive work and deep analyses and produced the kind of materials, the whole body of work, that Bob Avakian has produced. Show me anyone who even comes close.

Illusions of Freedom and Equality, the Reality of Dictatorship—And Moving Beyond Oppressive Divisions

AS continues: And the same goes for the stupid question—frankly, it is stupid: "Aren't we **all** leaders?" Well, it depends. What do you mean by leadership? And to what degree? We're certainly not all "equal" leaders. There may be a lot of people trying to provide leadership to a given process, in accordance with their different abilities and experience, but that's certainly not going to be all on the same footing, for all the reasons I just outlined. So when some people start saying things like "but, but...aren't we **all** leaders?" what they're really saying, once again, is "we don't want leadership, we don't acknowledge the value of leadership." It's just that typical pipe-dream of middle strata people who labor under the illusion that nobody's telling them what to do and nobody's leading them. Because one of the features of this kind of system, the capitalist-imperialist system, is that it has this **surface facade**, it wraps itself in a cloak of **bourgeois democracy**, which gives many people the impression that they are just existing and living in the society without being led. This system cloaks things, it hides behind a veil. Its core **essence** is one of exploitation and oppression for the many, both here and all around the world; but it tries to hide this behind a **surface appearance** that proclaims "certain rights and liberties," which are typically only extended to the few, but which can help them mask the true crushing nature of what lies at the heart of their system. The reality is that we already live under a dictatorship—it's the dictatorship of the bourgeoisie. But, the people who run this society don't tell you this, they don't want you to think about this. By contrast, the revolutionary communists are completely open and honest in explaining to people why we would all benefit from living under a **dictatorship of the proletariat**, organized and led in the right manner, a form and mechanism making it possible to lead things in the direction of abolishing those "4 Alls" we talked about earlier, getting beyond class divisions and class dictatorships altogether—all of which would benefit the vast majority of people. **That** kind of dictatorship would actually **be** in the interests of the vast majority of people, and its goal of getting to communism is a goal of emancipation for all of humanity. So communists don't need to hide this—they can be open and above-board about the need, and the positive character, of that dictatorship of the proletariat. But in current capitalist-imperialist society, in which we live under a **bourgeois** dictatorship, the people running things have no interest in telling everybody in school, and so on: "By the way, kiddies, we thought you should know that you live under the dictatorship of the bourgeoisie, and we want you to learn all about how we're leading you in accordance with **our** interests and objectives." [laughs] No, I don't think they want to do that. They'd rather give people this false sense of autonomy and freedom, and fill their heads with the idea that you can do whatever you want to do, that you can be whatever you want to be—and if you don't succeed, it's your own damn fault! As if they had nothing whatsoever to do with leading and enforcing the structuring of society in ways that continually restrict and distort people's lives, when they're not outright crushing them. So that's ridiculous as well.

As for the question of different ideas: Should everyone's ideas be given equal consideration? I don't know, do you really think so? Do you actually think all ideas are equal? How about somebody who says they ran into little green men from Mars the other day? That's an idea. Should it be given equal weight? What about the people who won't vaccinate their children, and who are creating a social problem because they are afraid of vaccinations, and they don't realize the value of vaccines? What about the idea that Caucasians are somehow superior to other races? Should that be given equal weight? I mean, there are all sorts of ideas, and many of them are wrong! Any individual will have tons of ideas. Do you really think they're all equal? Should we give equal weight and equal time in the schools and so forth to the people who, despite all the evidence, still don't believe in evolution, who believe the creation stories in the Bible, or the creation stories of any other religion, who imagine that some supernatural forces one day decided to...poof!...create life on earth, and, in particular, gave special place to human life...who believe that it was all done in just six days, and only a few thousand years ago, or any of the other ridiculous nonsense that is spoken to and taken apart in the *Evolution* book?

As a sidepoint here, I would like to once again encourage people to get into this book, *The Science of Evolution and the Myth of Creationism: Knowing What's Real and Why It Matters*, because I think it can really help people to learn not only the basic **facts** of evolution—to learn about the tremendous amount of concrete scientific evidence that shows how life on this planet came to be and how it changed, how it evolved, over billions of years, and how it continues to evolve—but to **also** learn about a lot of important points of philosophy and scientific **method**, which I think people would actually get a lot out of, and be able to apply more generally to all aspects of life, including to revolutionary processes. Forgive me for plugging that book here for a second, but I think it's actually worth doing.

But getting back to the argument that some people make that "all ideas are equal," should we really be giving equal time and equal weight to religious ideas of supernatural creation vs. the scientifically established theory of evolution, a set of ideas that has been repeatedly tested and verified in the actual world since the late 1800s, which is backed up by tons of concrete scientific **evidence**, and which has been **proven** to be true over and over and over again? I mean, come on!

Look, people can hold whatever idea they want to, they can have the idea in mind that they can fly off a roof by flapping their arms, as far as I'm concerned. Have whatever idea you want to, but don't try to tell me we should give all ideas equal weight! It gets back to the question of science. Ideas are not all equal, all ideas are not equally valid. And that's the point: Say whatever you want to say, have whatever idea you want to have, but don't tell me that they're equally valid, because all ideas are not equally backed up by concrete evidence, by scientific proof. Once again, science is an evidence-based process. And I can tell you, on the basis of science, that if you go on the roof and just flap your arms, you are **not** going to be able to fly, OK? [laughs] And that the outcome will not be very good. There is concrete evidence to back up **that** idea. Ideas should be evaluated not on the basis of what you happen to believe, or what some other people happen to believe, but on the basis of having been tested up against material reality, and found to be true, or false, on the basis of actual concrete material evidence. So if you have the idea that if you just flap your arms you'll be able to fly, that is a false idea that is not backed up by evidence. I'm obviously using a ridiculous example, but it's to make a serious point.

Question: That some of the ideas about why society is the way it is, and how you can transform it, are the equivalent of thinking you can jump off a building and fly?

AS: Yeah, pretty much!

A Fundamental Problem in the World Today: The Woeful Lack of Scientific Methods and Scientific Materialism

AS continues: One of the biggest problems we face today is that scientific methods and scientific materialism are in woefully short supply. People are not accustomed to really digging into the problems to see what are the underlying features of any kind of phenomenon or process, and what drives it. What *drives* the institutions of society under capitalism-imperialism? We talked earlier about the fundamental contradiction of capitalism—between socialized production and private capitalist appropriation—and the expression of that in the anarchy/organization contradiction and the driving force of anarchy in capitalism. There are rules to the game, so to speak. There are rules to any natural phenomena as well. **In any system, there are underlying contradictions that drive certain phenomena and that come to characterize them, and also come to provide a basis, a material basis, for these phenomena to change.** That's true if you're talking, for instance, about biological evolution in a population of plants or animals on the basis of the underlying contradictions represented by the underlying genetic variation in a population, or if you're talking about social systems, whose key underlying contradictions—those fundamental material "strains" that are built into the very foundations of a system and give it its distinguishing characteristics—provide the most fundamental **basis** for social change. This applies to the whole way the capitalist-imperialist system functions. Those great underlying contractions are what generates these horrible problems we've been talking about, but they are **also** what provides **the basis** to transform society in a revolutionary direction. Think of these core underlying contradictions of capitalism as the cause of a kind of great recurring "grinding of the gears" of their system—a problem they simply don't have the material basis to fix. They might squirt a little oil on the grinding gears every now and then, for instance in the form of a few concessions to the oppressed every now and then, but ultimately they are not capable of fundamentally reforming their system, there is simply not the material basis for them to resolve the great underlying contradictions of capitalism-imperialism while continuing to maintain and carry out the basic rules of functioning of capitalism-imperialism. Do you see what I mean? But if you don't go beneath the surface, if you don't dig into the deeper nature of the problem with some scientific methods to actually understand **how** things work in a society, and **why** they work the way they do, **why** the problems occur over and over and over again, and **what to do** about it, then you stay stuck on the surface, looking only at surface phenomena, and you come up with silly ideas, like maybe we can just convince a few capitalists to be more enlightened, or something.

It's like what I was saying about the environmental crisis. You can spend your whole life trying to convince people to compost their compostables and to use mass transportation as much as possible, and to walk more, and ride bicycles, and try to cut down on their use of fuels—you can try all sorts of things like that. But you're still on the surface, you're still making only tokenistic changes. So, even if there might be some value to some of these ideas (especially if they were being implemented in the context of a different system), under capitalism there simply is not the material underpinning, the material basis, for such ideas to truly take root, no material basis for them to really "stick," at least not with anything like the scope and scale of what's needed to really make a fundamental difference. So when you're talking about something like the environmental crisis, which is a global problem—talk about the necessary scope and scale!—little tokenistic transformations are just not going to cut it, they're just not going to get anywhere, because again, under the rules of the current system, there's just no material basis for them to really take root and change the terms of how we relate to the environment to a sufficient degree and in a way that is sufficiently different from what is going on today.

Question: Well, again, I think that, even though people wouldn't see it this way, one of those ideas about how to transform society that's the equivalent of people jumping off buildings and wanting to fly, is the idea, which holds a lot of currency, particularly among the middle strata, that the thing to do is just level everything off, let's just make everybody equal. And one of the really pronounced features of the world under capitalism-imperialism is incredible inequalities, both within U.S. society, and on a global scale: between different nations, between people of different nationalities, between men and women, between people who work with their hands and people who work with ideas. There are tremendous inequalities and divisions under this system. They're constantly generated. So I think even a lot of people who might be well-intentioned people look at this and say: "There are tremendous inequalities, don't we need to make everyone equal, don't we need to just level everything off?" But then that gets to the reality that, if you just declare that everybody is equal under a system that's constantly enforcing and generating tremendous inequalities, that's not gonna solve the problem. And then, even after the revolution, in the socialist society, if you just declare everybody's equal, but you haven't actually overcome those inequalities, and the basis for those inequalities, that's not gonna cut it either. So I wondered if you could talk a little bit more about this phenomenon—and I think this, again, is an outlook that comes from, or corresponds to the position of, the middle strata—where people say: Let's just have more democracy, or let's have "real" democracy, let's have everybody be equal.

AS: Well, that is a particular expression of the sort of petit bourgeois outlook that thinks you can...it's an idealist outlook which is, once again, completely devoid of scientific materialism, because it doesn't get into the question of what inequality is rooted in, and how could you actually overcome it. You can't just put a label on people and say, OK, everyone's now equal. What **generates** those actual inequalities that you just described? See, this lack of scientific materialism makes a lot of people turn away from the examination of what is fundamental in any society, what is at the base of any society, what are the forms of economic organization, the production relations. Any economy is seeking to meet the requirements of life, in a sense—but in what ways, by what means, and to what ends, to the benefit of which strata or sections of the people, and at the expense of which sections of the people?

This is why people do need to **learn at least some basics of political economy** to understand the connection between... this is why I was talking about the "4 Alls"...to understand the connection between class divisions and other oppressive divisions and inequalities, on the one hand, and, on the other hand, those underlying forms of economic organization and the production relations that characterize that economic organization, how the economy functions, and on what basis. And what all that has to do with the social relations, the social values, the way people relate to each other, the ideas people have. They are not disconnected! The way people relate to each other and the way they treat each other does not float in a vacuum, somehow separated from the underlying organization of society at an economic level and in terms of the production relations. Now, it would not be correct to say that all you have to do is set up a better underlying economy and somehow all the rest of it gets taken care of. That would be a narrow, economist view of things, which is why the revolutionary communists are always pointing to the fact that you have to continue to revolutionize these different spheres. If you have a socialist society, you're going to start by setting up a planned economy, in a way that is not geared to private profit—it's a whole different way of organizing production in the society. But if that's all you did, it wouldn't solve all the problems. You have to actually make concrete changes in all spheres of society, and there has to be struggle among the people to change the social relations and the ideas among people, and how people relate to each other. You have to work on overcoming those remaining differences in socialist societies, which are carry-overs from the old society: the differences between men and women, the divisions and conflicts between nationalities, and so on, the division between mental and manual labor, as you said. These are **remnants of the old ways**, of the old society, that will still affect the thinking of the people and the actual relations in concrete life among the people. So you have to consciously work at that, throughout the whole socialist transition period.

And not to go on and on about that right now, but I'm trying to underline the fact that if you don't have any materialism, if you don't make materialist, scientific analyses of the things that are underneath how a society is structured, then you can declare people to be equal as much as you want, or you can talk about democracy til you're blue in the face, but it has no meaning, because you haven't changed the foundations that keep regenerating all these oppressive differences on a daily basis. Something that's very relevant—I'd just like to read from Bob Avakian's book *BAsics*. This is *BAsics* 1:22, it's what's been called the "Three Sentences on Democracy," which I think are very relevant here:

> In a world marked by profound class divisions and social inequality, to talk about "democracy"—without talking about the *class nature* of that democracy and which class it serves—is meaningless, and worse. So long as society is divided into classes, there can be no "democracy for all"; one class or another will rule, and it will uphold and promote that kind of democracy which serves its interests and goals. The question is: *which class* will rule and whether its rule, and its system of democracy, will serve the *continuation*, or the eventual *abolition*, of class divisions and the corresponding relations of exploitation, oppression and inequality.

Again, this is from *BAsics* 1:22, and there is some complexity to that, but I think it would be very worthwhile for people...I don't think people are digging into that enough. It gets at the heart of some of these questions about what these class divisions are rooted in, what do different people, expressing different class outlooks, mean by democracy, whose interests are being served, and which direction is society moving in. As it says in *BAsics* 1:22, the question is: Does it serve the continuation of class divisions and the corresponding relations of exploitation and oppression and inequality, or is it a setup that will actually move towards the **abolition**, the eventual abolition of those divisions and those relations of exploitation, oppression and inequality? Which is it?

There is a material foundation, a material basis, in the way a society is set up, at the fundamental economic level. So, people need to think a lot more about the connection between ideas and the material underpinnings of things that can give those ideas more or less validity, and more or less of a basis to actually take root in the real world, as opposed to just floating out there in a vacuum, as just some abstract ideas. You have to connect ideas to material reality. Something can be a good idea, maybe somebody a few centuries ago had a good idea about how to remake society, but there wasn't the material basis at that time to do it in line with the ideas that they might have had at that time. See, there's a connection. And we live at a time in human history where there actually is a material basis for this new set of ideas, corresponding to new relations and new directions in society, to actually take hold in the real material world. But it requires sweeping away the existing system and the existing structures, in order to clear the ground to be able to start doing that.

Question: Yes, a really important point here, which I actually meant to bring up before, in my last question, is this: At the heart of the communist revolution is completely overcoming these profound inequalities that are generated under capitalism-imperialism, but you can't do that just by declaring that those inequalities are over and declaring everybody equal. You actually need a revolution to sweep away this system that's responsible for it. Nor, after the revolution, in the socialist society, can you just declare everybody's equal. There's a whole process, as you were saying, of overcoming these inequalities. Obviously, getting into these questions in much greater depth is beyond the scope of this interview. But, off of what we've been talking about, I just wanted to mention a couple of other sources which would be good to refer people to, to get deeper into some of these questions. There's the Constitution that you mentioned before, the *Constitution for the New Socialist Republic in North America (Draft Proposal)*, which, on the basis of Bob Avakian's new synthesis of communism, lays out a very concrete and sweeping vision of what the future socialist society would look like, and gives you a sense of the whole process that would be involved in overcoming all these inequalities. There's Avakian's work *Birds Cannot Give Birth to Crocodiles, But Humanity Can Soar Beyond the Horizon*, which people can find at revcom.us, and I would direct people to that. And I know that there's just been a reprinting, a republishing in India, of Avakian's work *Democracy: Can't We Do Better Than That?*, which is also very relevant in relation to all of this. So I just wanted to highlight those works, because, again, to go deeply into these questions is obviously beyond the scope of this interview and what we can do now.

AS: I also think Avakian's work *Communism and Jeffersonian Democracy* would be a very good resource for people who want to dig into these questions.

Question: Yes.

AS: I also want to point people to another concept which I think is important in relation to this question: Some people ask, "Shouldn't our goal be democracy?" But it keeps coming back to this: You're going to go off track if you don't actually try to apply scientific methods and look for the underlying material reality underneath ideas. Again, you can't have these ideas floating in a vacuum. And one thing I've always found very provocative is this quote from Marx, which I can only kind of paraphrase right now, where Marx, using broad characterizations, talked about what the shopkeeper and the democratic intellectual have in common. He talked about how, in their daily conditions of life, the two are as far apart as heaven and earth. A shopkeeper and an intellectual, in terms of what they actually do, and the kinds of things they think about, on a day-to-day basis, they can be as far apart as heaven and earth. But what they have *in common* is their frequent inability to see beyond the narrow horizon of bourgeois right, to envision and aspire to a radically different framework for the organization of society. And, from another angle, they all too often end up working on their bottom line, in the case of the shopkeepers, or seeking petty advantage or advancement in the case of the intellectuals. But whatever form it takes, they're basically reinforcing their social status and position under the existing framework, they're not actually striving to break out of the existing framework of the current organization of society. And, on the surface, they might seem to be very different components of the middle strata, but actually this is where they often come together and have something in common: they're not breaking out of this established framework.

You know, I don't fault people for being unfamiliar with a radically different framework that they've never encountered before. It's almost like going to another planet! You have to really dig into it, to even learn what it would look like, what it would feel like, what life would be like on a day-to-day basis. So, I don't fault people for not automatically knowing about the new framework and what a socialist society based on the new synthesis would look like. But I do fault people for not attempting to learn about it, when somebody's done all this work to actually develop the framework. And you can get a very concrete feel, including in the *Constitution for the New Socialist Republic*, for what it would look like. It's very concretely laid out. There would be many things that would be changed overnight that most people would really welcome. For instance, you wouldn't have police gunning people down on the streets. You wouldn't have women unable to go about freely, especially at night, because of fear of being raped, or being battered by their spouses and having no recourse. You wouldn't have these wars for imperialist expansion that people get swept into. You wouldn't have an insane way of structuring society that, on a daily basis, is driving humanity closer to extinction through environmental degradation.

You'd still have lots of problems, in all these different spheres, and you'd have to keep working on them. You'd still have to work a lot on transforming the thinking of the people. But you would have **a material grounding** that would give you **a basis** for doing all this, in a systematic and ongoing way, but also in a way where people would feel like they could have some ease of mind, they could have some differences and wrangle over questions, and there would be air to breathe throughout the society. This, again, is a hallmark of the new synthesis. These changes would not be accomplished with jackboots and turning everybody into robots, thinking like automatons—not that that was the case in previous socialist societies either. That would be a slander. But to a much greater extent under the guidance of Bob Avakian's new synthesis of communism, there would be an appreciation for the process of wrangling, of ferment, of dissent, of people wrestling with ideas, experimenting in various ways with how to transform things in an ongoing way.

Which Side Are You On?

Question: In what you have been saying, there is some really illuminating and important discussion in terms of how different ways of thinking relate to different class positions in society, and how that is bound up with this snark and slander, and personal attacks, and opposition and dismissal without engagement, directed at Bob Avakian. There's a lot that's really important to dig into, in what you've been saying about that and the deeper foundations underneath that. But I wanted to return to the question of the harm this does. There are people who are more outright, straight-up, counter-revolutionaries whose defining purpose is to tear all this down. And then there are people who may not be under that umbrella, so to speak, but they're still engaging in this pettiness, nastiness and snark. So, I wondered if you could say more about the implications of that, the harm that does?

AS: Well, in the '60s, people used to commonly challenge each other by asking: "Which side are you on?" And one of the things to realize is that some of these people today who constantly engage in snark and slander attacks, besides being petty and nasty, they are really uninspiring and devoid of much substance, they don't put forward any substantial answers themselves. They are very annoying. Now, I do think **it's important not to get diverted or overly preoccupied** by such phenomena. It's just part of the deal, there are going to be people doing this kind of nipping at your heels when you're trying to do good things. It's just part of the deal, and you really shouldn't overly worry about it and get overly preoccupied with this, on the one hand. For one thing, all this will likely start to change as broader masses of people—basic people and people from other strata who are actually striving to come together and unite to make a better world—start to take more initiative and start telling these people: "Go take a hike, we're sick and tired of your attitudes and your getting in the way—just get out of our face, we don't want to put up with this!"

On the other hand, in terms of the objective harm that this does, I don't want to overblow it, or get overly preoccupied with it, but it is important to realize that it's not just an annoyance. When people are trying to make a revolution in a country like the U.S., there's a whole process, a whole extended process of political preparation before you can get to the point where the conditions emerge and you have what's needed to go for an actual revolution, an actual seizure of power, which is what you're working towards. But you can't concoct that overnight. The conditions have to emerge, and the conditions have to be created, preparing the terrain, preparing the people, and preparing the vanguard party—**all** of that has to be systematically worked on and prepared, over a fairly extensive period of time. And, as you're doing that, as you're working on that, you can be sure that the people running society are not going to just sit there and ignore you. If they think you're having absolutely no effect, and absolutely nobody is interested in what you're doing, well then, they might ignore you for a while. But they minimally keep an eye on you, because they recognize that, in fundamental terms, you are a threat to their system. And especially if you start to broaden your ranks and connect with more and more people and have broader and broader influence...they start to realize...look, they have their own analysts, people who are perfectly capable of analyzing the fact that there is substance in the leadership of someone like BA, and that this has a real potential to cause problems for them. And so they don't just sit on their hands. They take all sorts of steps to try to attack and destroy a revolutionary movement as it's gaining strength.

There's a whole history of government repression and COINTELPRO activity aimed at BA and other revolutionaries who came forward during the upsurge of the 1960s. People should read BA's memoir, *From Ike to Mao and Beyond: My Journey from Mainstream America to Revolutionary Communist*, and people can also look at the *Timeline* of BA's political activity and revolutionary leadership since the 1960s. You'll be able to see the kind of repressive moves that were made against him and some other revolutionaries at different junctures. You can learn about the not-so-veiled threats that have come from a number of different directions, and the ways some of the government intelligence services actually encourage and promote campaigns of slander against revolutionaries, and especially revolutionary leaders, as a way to discredit them and to try to turn people against them.

That's not new. That's standard operating practice for agents of the government. They plant slanderous stories, they encourage people to repeat slanders and gossip, as part of undermining things. But they also assassinate revolutionary leaders. And when they're not assassinating leaders, they are still doing a lot of other things to disrupt the revolutionary movements, and to make it difficult for revolutionary leaders to do their work, to actually have the kind of conditions where they are able to keep working. So you're talking about a very antagonistic relationship, objectively. There are times when the repression and the concrete moves—legal charges, hounding, physical threats coming from government authorities, and so on—there are periods when all this is very acute and carried out in a very blatant way. And there are times when repressive and disruptive moves are carried out a bit more in the background, in a way that is a little less overt and where the government's hand remains a bit more hidden. Historically this generally has to do with how authorities are evaluating the degree of threat posed by revolutionaries at a given time or in a given period. And it can also have to do with some differences and disagreements they can have among themselves about what is the best way to contain or suppress revolutionary movements and leaders at a particular time—ignore them? slander them? slam them with legal charges? carry out targeted assassination?—including at times when some of the ruling class functionaries might be arguing that they have a lot of other fish to fry and a lot of other problems to deal with, on the world scale and not just in this particular country. So their attentions can be divided, at least for a time. But it's important to consistently keep in mind that, objectively, it is **always** the case that genuine revolutionaries are in **a fundamentally antagonistic relationship** with the other side, with the people who are running society. You are, after all, trying to build a revolutionary movement that will actually involve and organize broader and broader sections of people, training thousands to eventually be able to lead millions, all in order to get to the point where it will be possible to actually overthrow their capitalist system, dismantle their state institutions, and begin to build up a whole new society on the basis of completely new organs of state power and other institutions. So yeah, this is objectively an antagonistic relationship, OK? [laughs] They're not just gonna sit there and passively wait for you to get to the point where you can do that. So, that's the constant threat in the background—and, again, they're perfectly capable of recognizing the particular qualities of a revolutionary leader of a certain kind, of somebody who actually has the experience, skills and overall abilities to lead the whole thing. They talk about, and some of them study, people like a Lenin or a Mao. They understand that not every member of the revolutionary party, or every person in the revolutionary movement, in those societies was on that level. They understand the particular and crucial role of developed and advanced leaders.

So, as I'm reminding people of all this, think about this as the larger context in which some of the "haters" and "snark culture people" are operating. These people spend a whole lot of time and energy spreading slander and vilifying serious revolutionaries and revolutionary movements, and in particular stand-out revolutionary leaders like BA. Sometimes you have to wonder why they don't have anything better to do with their time! They often hide behind their computer terminals to launch the most vile attacks to try to discredit and diminish BA, distort his works and his basic positions, all in an effort to try to divert and discourage people from learning more about him and following his leadership. If they don't agree with BA, why don't they just concentrate on doing their own work on whatever issues concern them in society and the world, rather than obsessing about tearing down BA? The reality is that they would like nothing more than to see *you* **not** look into, **not** study, **not** discuss and **not** debate what BA has to say. Sometimes it's frankly hard to tell the difference between an actual pig and a "pig-like" hater or political opportunist in the broader society. They're probably intertwined to some extent.

Another thing these people are doing these days is working to drive a wedge...to undermine support for the revolutionaries on the part of people who are not themselves communists or even revolutionaries but who nevertheless would like to contribute their time, their energy, their money or other resources to support the work the revolutionaries are doing to develop resistance to the outrageous social abuses of today and to encourage broad societal discussion and debate about what BA has to say and about the problems of society and different possible ways of going forward.

Some of these "haters" *literally* go around to people who have been inclined to provide such broad-minded support, for these types of reasons, and tell them to stay away from those revolutionaries, to turn on them, to publicly denounce BA, to not even read, or discuss with others, BA's books, films of his talks or other works, to publicly denounce him, and above all not to donate any of their money or other resources to these revolutionaries. This is really low-down and disgusting, right? I mean, of course revolutionaries need support in various forms, including donations of lots of money to spread their message and their books, films and other materials very broadly throughout society. A revolution's never gonna happen without this kind of broad support. The revolutionaries themselves are obviously never going to get rich by trying to spread revolution in a country like the U.S.—if your objective is to get rich as an individual, find another line of work! [laughs] And the revolutionaries are also obviously not going to get big government grants or grants from the big private foundations to help them do their work! So there's never going to be enough money. As in everything else, the revolutionaries have no alternative but to rely on the people to help advance and spread the revolution. That's why it's so important when people who have very little money of their own organize things like bake sales in the housing projects because they know, in their heart of hearts, that this is important to their lives and must be supported. And that's why it's also so important when some more privileged people, who do have some money but who also have a moral conscience and sense of social responsibility—people who may not even be convinced of the need for revolution, or of this type of revolution, or of BA's qualifications to lead—nevertheless decide to donate significant funds because they realize that this is a serious and well-intentioned project, one that can minimally promote widespread and much needed discussion and debate of critical issues broadly throughout society, that is building up concrete resistance to the outrages of today, that is broad and generous-minded in its outlook and approach and concretely working to unite a very broad variety of people in the process. Such supporters in particular are being targeted by the "haters" and merchants of snark. Hopefully, a lot of them will think for themselves and will increasingly be able to see through these kinds of unprincipled attacks, stand firm on the basis of their own integrity, ask lots of pointed questions, dig into things more deeply themselves rather than swallowing slanders uncritically and turning against BA and the other revolutionaries. Unfortunately, principle and integrity seems to be in very short supply these days, but it is precisely on the basis of personal principle and integrity that many should find it in themselves to resist the snark, and to maintain, and even extend, their support of the revolutionaries.

People should reflect on this more. The lies and distortions about BA and his works which are spread by people who generally tend to "dismiss without serious engagement"—who have never really seriously cracked a book or dug into his whole extensive body of work, who would not be capable of engaging in any kind of really serious conversation or substantial analysis and critique of his method and approach and his new synthesis of communism, but who are nevertheless only too willing and anxious to get other people to distance themselves from this revolutionary leader—all this plays right into the hands of the enemy, plays right into the hands of the capitalist rulers. It doesn't really matter whether these people are doing it consciously or not, whether or not they themselves have thought this all through: whether people are doing it consciously or not, **objectively** it's a form of **complicity**, of being complicit with the enemy. So, again, I would challenge these people: "Is that really what you want to do? Do you really want to play the role of being complicit in this way? Stop it! Just stop doing this. This isn't a parlor-game. What you're doing is disgusting. What you're doing is unconscionable. What you're doing is objectively being complicit with the oppressors. So just stop it! If you have differences of substance, that's fine, by all means write them up, discuss them, engage in substantive and principled debate. But stop vilifying and slandering and trying to turn people away, stop making that a focus of what you do in life, because you're just ending up on the wrong side of history here. You're objectively doing the enemy's work for them."

Question: And it also seems that people more broadly who are not themselves doing that, vilifying or engaging in snark and slander—people who are politically active or politically awakening or just people in society more broadly—need to learn to very sharply draw the distinction that you just drew, between principled disagreements over content versus snark and slander and unprincipled attacks. There actually needs to be the kind of standards where people draw those distinctions, and know how to draw them.

AS: Yes, that's true, that's an important point. Regular people, if they come around, they start to check things out, and they come across some slanders against the Revolutionary Communist Party, and in particular this vitriolic vilification and personal attacks against Bob Avakian. I think the correct response should be something like this:

> I don't want to hear it! This is a guy who's dedicated his whole life to this. He's trying to put forward an analysis of why we have all these deep problems in society and all these great injustices, and how we could have a whole new society. Maybe I don't know if I agree with him yet. Maybe I don't know if I think we need a revolution or not. Maybe I don't know if I agree with the vision he's bringing forward. I just don't know enough about it yet, but I'm planning to get into it. What I do know is that this is the kind of thing we should be talking about, these are the big questions we should be discussing and debating, to either express agreement or disagreement, and to further explore these questions. That's what people should do. So I don't want to hear these stories, this gossip, these attacks, or whatever. This is disgusting and it doesn't serve anything good. It's just trying to tear somebody down. For what? When someone's done a lot of work and he's putting forward a way to get out of the mess that society is in, I want to hear more about **that**. I don't know yet if I'll agree or won't agree, but I want to dig into the substance of it.

Come On People! There's a Place for You

Question: Well, just to come back to the question of the middle strata, I wanted to ask: What would your message be to people who may be from the middle strata, or who might be influenced by the outlook of the middle strata that we were talking about, but who are not firmly entrenched in the camp of opposition, whether they're professors, artists, intellectuals, etc.? What would your message be to those people?

AS: Well, to put it simply, **the revolution needs you**. Get with it. Join in. Follow your conscience and your convictions. The first thing is, you have to care. If you **care** enough about what's going on...If you **care** enough whenever you hear of yet another unarmed Black or Brown youth being gunned down by the police... If you **care** enough about a whole people being incarcerated in unprecedented numbers in this country...If you **care** enough when women are dragged through the dirt and dehumanized and degraded and reduced to practically chattel slavery...If you **care** enough about the destruction of the global environment...If you **care** enough about the fact that these destructive wars are being waged all over the world, and **in your name**...If you **care** enough that whole groups of human beings are being declared "illegal" and hounded, harassed, imprisoned and even murdered just for seeking a better life, and to escape desperate conditions that have actually been created by the system of capitalism-imperialism. If you care enough about all these outrages and the intolerable conditions of the people—if you care enough, even if you yourself are having a pretty good life, and are not suffering so directly from this, but you care enough about the fact that this goes on continually and on such a large scale—then that's the essential first ingredient. And if you do really care, then come forward! Dig into the analysis that has been worked out, in particular by Bob Avakian, about the **source** of these problems. Go deeply into it. Study the materials. Discuss it with your friends and family and people more broadly. Challenge what you don't understand or what you don't agree with. Bring your ideas into play. Argue about it. What you *do* agree with, help promote it, help spread it. In fact, help spread it even when you don't agree with it, to help to make it a subject of broad discussion and debate society-wide, for the reasons I outlined before about how important it is to have these kinds of substantial discussions broadly in society.

Now, I don't want to leave this interview having given the impression that I think that all the middle class people are really messed up and have no place in the revolution. That would be completely wrong, and it's not at all what I'm saying. I will say that I am very frustrated with a lot of middle class people these days who seem to approach things very superficially, have very low standards, often proceed primarily from "self" and don't seem to care enough to get with it. I want to challenge people—in particular the youth of these strata—to wake up and get with it. Do some work, dig into things. You have a great deal to contribute. You have a great deal to bring to the process, in terms of ideas and abilities. We're not talking about a revolution that would exclude you. We're talking about a revolution where you should definitely be part of the process, and part of building the new society. And there is definitely a need to bring together these very diverse strata of the people: people from the basic masses, from the inner cities, with students and professors and artists and intellectuals and scientists.

And hey you, you scientists out there! The people working in the natural sciences, **where are you, my people**? [laughs] Come forward! You should be *thrilled* that someone has emerged who is actually applying *scientific methods*, consistently and rigorously, to the problem of how to change society in a more just and righteous direction that would benefit the majority of humanity. This should be right up your alley!

Scientific methods are not to be restricted to just working with the natural world. Human society is part of material reality, like everything else. It's matter in motion, just as much as anything in the natural world. So if we want to change it in a better direction, we should apply scientific methods, and not just superficial subjective notions or relativistic ideas about truth as some kind of competing narratives, or proceeding from identity politics. Many people these days in universities and elsewhere reject the idea that truth is objective and can be determined though an evidence-based scientific process; they argue that one subjective "narrative" is just as valid as another, and they think that the way to approach tackling societal problems and social divisions is for everyone to proceed from "identity politics," the idea of promoting and proceeding from the supposed interests of just one component of society, from one or another "identity," such as "just women," or "just people of color," or "just people in the Third World," or whatever. We should really not be getting pitted against each other and should not be dividing out into such narrow camps in that way.

There are deeper material underpinnings to the biggest problems of society and the world that need to be explored systematically, including in order to be able to detect their common roots. Come on, people! Use some actual scientific methods, some rigorous scientific methods, to dig into the deeper nature of the problems and to explore and evaluate possible pathways and solutions to make for a better world for all. Join in with **that**. You will be welcomed in this revolutionary process, and, if you actually deign to dig into it with substance, I suspect you will find a lot to like in Bob Avakian and his methods and approach, his seriousness, his rigor, as well as his sense of humor and liveliness, and the kind of person he is, which actually is a good indication, in and of itself, of the kind of society that he is envisioning and concretely striving to bring into being.

Question: And it also seems like something really key is people not being afraid—the people that you're appealing to right now—not being afraid to step outside their comfort zone, to have their assumptions challenged, to have a certain class outlook that is influencing them be challenged, being willing to go there.

AS: Yes, I agree, that's really important. Come on! What's the worst that can happen? That you'll find out that you were looking at things the wrong way for a long time? Discovery should be inspiring. You know, one of the things that natural scientists often get really excited about is the discovery of an error, or an anomaly, or something that is revealed to have been wrong in the way they had been looking at things. Why is this so exciting? It's not because anybody particularly likes to feel like they've been wrong for a long period of time. [laughs] It's because it's very exciting, certainly to most natural scientists, to discover the actual truth about something, the actual evidence about some aspect of reality, even if it's not at all what you expected to find and doesn't at all correspond to what you had been thinking up to that point. It opens up whole new dimensions. You learn from it, including by figuring out **why** you had previously been wrong—not just that you were wrong, but **why** were you wrong—and following through on the implications of all that to gain even further insights. So, for instance, with a scientific approach to the social sphere, you might discover that you'd been holding a lot of unexamined misconceptions about what's really involved in socialism and communism, about what had actually characterized the first wave of socialist revolutions, in the Soviet Union and China, about the nature of the challenges they faced, and about both the actual advances and the shortcomings that had been involved in these first attempts to develop socialist societies. And you might be surprised by what you discover if you take a consistently evidence-based approach to digging into all this.

You might have been "told" or taught in school that these societies were horrible, that all this was just a terrible disaster, that everyone suffered tremendously, and so on. But you probably never really looked into it very deeply, because these were just things that you **assumed** were true, things that supposedly "everybody knows," and you relied on this general populist consensus (as well as the propaganda of the people running things) rather than employing a more systematic and critical scientific approach to uncovering the actual patterns of evidence. But you know, this is really not acceptable. There is no intellectual integrity to this kind of approach. You can't just blindly repeat that you've "always heard" and that "everybody knows," so it must be true, that the first socialist societies were a horror, that Mao was a monster, and so on. What happened to critical thinking? What happened to social responsibility among intellectuals? What happened to actually doing the work to systematically dig into patterns of accumulated evidence as a way of uncovering what's true?

So maybe now, if you're a person with even a modicum of intellectual integrity, you decide to do some work to actually learn more about all this: You could go to the *Setting the Record Straight* project, and you could go to the revcom.us website, you could read the special issue of *Revolution* with the interview with Raymond Lotta, *You Don't Know What You Think You "Know" About...The Communist Revolution and the REAL Path to Emancipation: Its History and Our Future*. You could dig into this whole experience of the first socialist societies, and you could actually dig into the references and research materials and look at the actual evidence—what people were trying to do, what they actually accomplished, what their shortcomings were—and sort out truth from fiction in this way. Maybe then you would find yourself concluding, "Well, I have to admit I had a completely different view of this, I was off on this, I realize now I had some real misconceptions about all this." Well, that wouldn't be a disaster, right? It's not going to kill you. In fact, it's likely to open new doors for you, to challenge you to explore things further and even more deeply, and maybe you'll even find yourself motivated to contribute in your own ways to trying to advance humanity in some new directions, to break out of the currently stifling and oppressive frameworks.

Sometimes people from the intellectual strata in particular, they worry that the socialist revolution and going in the direction of communism would suck all the life out of things, especially for people such as themselves—they worry it would impose a lot of narrowing restrictions on them and their work, and in general suck the life out of the life of the mind, the world of ideas, the world of experimentation and innovation, and so on. But that, too, is slander and misconception. Just read Bob Avakian's works, OK? You'll get a sense, a deeper sense, of the kind of vibrant society that is being envisioned, and what it would take to create the material basis to actually give large scope, expansive scope, to intellectual curiosity, experimentation, investigation, discovery, in a way that ultimately does serve humanity, but not in a way that is restrictive and stifling; in a way that allows for going in a lot of different directions, for pursuing projects—research projects, artistic projects, and so on—that would not necessarily bear fruit in any kind of narrow or immediate sense...you might not know ahead of time exactly where a particular project is going, or where it might end up. It might not lead to anything, it might just be some kind of dead-end. But that's all part of the process of advancing science and art and culture more generally. With Bob Avakian's new synthesis of communism, you actually would have the ability to experiment and explore widely, without being overly tied or restricted to just the production of immediate and palpable results, narrowly conceived. This would be **anything but** a stifling and suffocating society. You artists and scientists and other intellectuals, you should actually be dancing in the streets at the idea of what is being envisioned here. And think about what a contrast this poses, what an inspiring contrast this poses, to the society we live in today!

Communist Leadership, Not a Condescending Savior

Question: Before ending, I wanted to come back to the question of the role of BA's leadership in your own development as a revolutionary and a communist. Maybe you could talk a little bit more about that.

AS: Well, as I said, it became kind of clear to me very early on, even in the 1970s, that there were some very special characteristics to BA's methods and approaches. Some of the work that he was developing even back then just didn't seem to be like what more typically prevailed in the movements of the times. It was more serious, it was more substantial. It was also more nuanced and had an appreciation of complexity. And there was a very important way that theory was being combined with practice. It was not armchair Marxism. It was definitely tied to the practical movements of the day, actually trying to concretely advance the revolutionary struggles in the direction of an actual revolution. It was very concrete that way, as it continues to be today. But it was already, back then—as is still evident today—very much characterized by deep theory and deep development of analysis, and it stood out in contrast to many kinds of superficial and limiting approaches to social change, along the lines of the view that "the movement is everything, the final aim nothing," or various forms of economism, approaches aimed only at bringing about a few minor improvements in the situation of the oppressed while ultimately remaining squarely within the bounds of this horrific system. BA was not tinkering with superficial reforms, he was working on recasting the whole framework at a more fundamental level, and this definitely appealed to me.

When I think of BA's theoretical and practical guidance all along the way, over more than four decades, I also think about key junctures along the way. I won't do justice to all of them right now, but just to give some examples. From early on, there was deep analysis of the Black national question in the United States. Even way back then, that was very formative for me, actually. I thought: This is the only really scientific, substantial analysis of the national question that I've seen that actually points things in the right direction. There was deep and historically grounded analysis, and there were polemics, on the one hand talking about why there needed to be a real revolution, but also warning against the dangers of adventurism, of just a small group going off trying to make a revolution without the conditions and the masses of people that would make a revolution possible, and how irresponsible that adventurism would be, how it would actually not lead to a real revolution. There were also important polemics all along the way, from an early time, about the problems of economism in the communist movement, and about the historical problems of chauvinism and nationalism in the international movements. These were extremely important polemics. I think BA single-handedly brought forward and reinvigorated a lot of the best of what Lenin was about, his critique of economism, for instance; and BA actually has deepened that analysis and taken it even further.

There has been the deep analysis of the experience of the first wave of socialist revolution, including the whole experience of China, and deep analysis of the restoration of capitalism in China, which a lot of revolutionaries at that time were not even recognizing **had** taken place, or not really understanding **why** it had taken place. There was a dissection of the different lines within the Chinese Communist Party, and a materialist analysis put forward of the underlying **basis** for these lines to take hold within the overall class struggle in China, focusing on the class contradictions in socialist society and the underlying material basis for this, what the problems were that the Chinese revolution was encountering, once it had entered the stage of socialism and was seeking to advance on the road of socialism, and also bringing to the fore some of the shortcomings in the approach of the Chinese communists that had actually ended up making the society more vulnerable to capitalist restoration, even though, as Bob Avakian also analyzed, the shortcomings and mistakes of the communists were not the **main** reason that capitalism was restored. So, there was crucial analysis at critical junctures like that.

Then, with the waning of the movements of the 1960s in this country and internationally, and the sharpening of the contradictions between the imperialist bloc headed by the U.S. and the rival imperialist bloc headed by the Soviet Union (which already was no longer socialist, despite making claims to the contrary up through the 1980s), there was analysis of the danger of world war. When later there was a dramatic turn in world events, with the Soviet Union and its bloc unraveling and openly becoming capitalist, this was analyzed as well, with Bob Avakian's leadership, and this included analysis of some secondary errors in the RCP's approach, which had tended to be somewhat mechanical and one-sided about the danger of world war, a danger which was real but was overstated somewhat in the RCP's initial analysis. If you look at *Notes on Political Economy*, published by the RCP in the 1990s, you will see both important analysis of the dynamics of the world imperialist system and its capitalist foundation, and important summation of both the mainly correct analysis of the RCP regarding the danger of world war during the 1970s and '80s **and** of the secondary errors, of method as well as specific content, in that analysis. There is a great deal to learn from both parts of this. It goes back to what I was saying earlier about the scientific method, and how an important part of that, and something that is welcomed by this scientific method, is coming to recognize and learn from mistakes you make.

There have been other very important contributions to the strategic orientation and approach to carrying out the revolutionary struggle, not only within a particular country but throughout the world. One of the key contributions of Bob Avakian has been his deepening of the understanding of the basis for and crucial importance of **internationalism**, captured in the basic orientation that the whole world comes first. This involved analyzing and critiquing some of the misdirections of some of the revolutionary organizations and parties in different parts of the world that went off track, particularly in the direction of replacing communism with nationalism. It's not enough just to say that they went off track—it's important to get into **why**. Otherwise, you don't learn anything from it. And you can't do better when the opportunities arise again. Again, I would urge people to read the polemic by the OCR of Mexico on this question, "Communism or Nationalism?" and other important polemics in the online journal *Demarcations,* which can also be accessed through the revcom.us website.

And during this whole period, there has been very important analysis by Bob Avakian of the need and importance of consistently doing revolutionary work in a non-revolutionary situation, and what it means to carry out this work, to prepare for a revolutionary situation.

Again, these are just some examples, and in this interview I cannot do justice to the whole scope of Bob Avakian's body of work and to the crucial contributions and breakthroughs I feel he has made, in terms of developing the theory and strategic approach for a new stage of the communist revolution in the world.

And within the Revolutionary Communist Party itself...and this has been made public, so people can read about it...Bob Avakian initiated and led a Cultural Revolution right within the Revolutionary Communist Party, as a means and method for dealing with seriously revisionist lines and tendencies that were threatening to take the Party fundamentally off track, in the direction of economism, reformism and accommodation to the imperialist system.

With Bob Avakian's leadership, there have also been sharp and profound critiques of identity politics—of nationalism, and identity politics around the woman question—and excavations of the philosophical basis of all this in unscientific idealism and **relativism**, in the notion that there is no objective reality, or that we cannot really know objective reality, and there are only different particular "truths," or different "narratives" of different groups or individuals, as I was talking about earlier. BA has also made a very important critique of "revenge lines" among the basic masses, emphasizing, in opposition to this, that the communist revolution is **not about revenge** but about the **emancipation of humanity**, the ending of **all** relations of oppression and exploitation.

Around many major social questions, he has deeply probed and investigated, and brought forward many important analyses, but it's always been harnessed to the strategic objective of communist revolution and emancipating humanity. His orientation has consistently been: Why are we doing this, why are we even bothering to wrestle with any of these questions? It's not because it's simply interesting, or just so we can have stimulating conversations with each other and reflect on how clever we are, right? [laughs] We do this because we're actually trying to make a better world. So, what do all these questions have to do with **that**, with actually making a better world? In BA's approach, the questions explored and wrangled with have always been harnessed and directed like that. I've always appreciated that about BA.

And another hallmark of BA is that there's always been a real willingness to struggle with people honestly and forthrightly about the obstacles in their own thinking. He has no hesitation about struggling very honestly and forcefully with people of every strata, including people at the base of society. You're never going to get any kind of condescending savior attitude from BA! [laughs] He's going to go right down on the ground with you and tell you what he thinks. He'll tell you straight up what he thinks you're doing wrong, or thinking wrong, **and why**. He's not going to pull punches, or be in any way reluctant to struggle with you, just because you might be from an oppressed grouping or something. **He will respect you enough to struggle with you.** And I really appreciate that attitude as well.

A Consistently Scientific Method and Approach

AS continues: What wraps it all together is, once again, the question of **method and approach**. What matters the most to me is BA's method and approach, and in particular the philosophical-methodological ruptures he's made relative to the previous experience of the communist movement. He's taken things a lot further in terms of putting communism on a more consistently scientific basis, even correctly recognizing that, like any genuinely scientific theory, communism as a science must necessarily be falsifiable. Unlike a religion, a genuinely scientific and comprehensive theory must be open to being subjected to repeated testing up against the reality of the material world and to the risk of being ultimately disproved and rejected, but only on the basis of systematically acquired concrete material evidence, accumulated over a whole period of time and coming from a multiplicity of directions. That's why we can say, for instance, with a great deal of confidence, that the theory of evolution is necessarily falsifiable (as it must be, to qualify as a scientific theory) but that it has **never**, in the more than 150 years now since Darwin, been **falsified**. Despite all the best attempts of religious forces to disprove it, the theory of evolution has only been repeatedly strengthened and reinforced over time, by countless instances of concrete evidence of its validity. Communism as a science must be looked at the same way.

Its core principles and analyses are grounded in the material world and are being, and will be, tested, repeatedly, up against the actual workings of material reality. To say that, like any good scientific theory, it is "falsifiable" does not mean that it will end up being falsified: It simply means that it's open to critique and multi-faceted investigation, and to being repeatedly tested in the concrete up against actual material reality. As a scientist, as somebody trained in the sciences, that's incredibly important to me: to understand that this is getting away from approaching communism with any kind of "religiosity," and really looking at it as a scientific process that is grappling with actual material reality as it is, and that is willing to recognize mistakes, and misdirections—your own mistakes, as well as the mistakes of others, historically or in recent history. To recognize these mistakes and see them as part of a process, to dig into them and what you can learn from this, so that you actually come out the other end being able to advance things on a better basis that is more in correspondence with reality, more deeply in correspondence with actual reality—that kind of scientific method is very much in contrast with the assorted populist epistemologies which tend to define truth in relation to how many people think something is true, or which section of the oppressed people think something is true, and so on. BA doesn't go for any of that, you know. He just goes wherever the evidence takes him, on the basis of serious and deep analysis. He is willing to confront that, even when it means confronting errors, or, as he has put it, confronting the truths that make you cringe. That's the hallmark of a good scientist. And, more than anything else, that method and approach is what really distinguishes him, really makes him stand out, and it is something people should very consciously study and learn from, and take up themselves.

I feel that all these things I've described, especially this rigorous, scientific method and approach, have been a source of continual inspiration and concrete training for me personally. I may have had scientific training in the natural sciences, but it's a whole other deal to really apply this kind of understanding to the analysis of society, and to the transformation of society. There are so many ways you can go off track! For instance, you can fall into subjectivity around certain questions. I care passionately about the woman question, the oppression of women. And one of the questions that was posed in the '60s was: Could you actually fight for the emancipation of women *within* the revolutionary movements, and *within* the movements for socialism and communism, or did it have to be taken up as a completely separate process? I had a certain level of understanding of why I felt it should be taken up as part of a single, overall, more all-encompassing process. But I would say that, in an ongoing way, my thinking about such questions has been further enriched by delving into BA's whole overall method and approach, and I feel that all this has really helped sharpen my thinking and my work on the question of the oppression of women, and on what kinds of fundamental societal transformations would have to be instituted in order to actually achieve widespread emancipation and not get pulled back repeatedly into the more limited and limiting framework of identity politics, with its strikingly narrow conception of the roots and ongoing features of women's oppression and how all this relates to the overall and fundamental relations in society which are all in need of a completely radical and systemic overhaul.

And this is true around many different questions. One of the main things that's influenced me, in terms of BA's leadership, is the scientific rigor, the consistency, the willingness and ability to proceed methodically and systematically. If you're spontaneously being pulled by some sentiments in society, or by your own particular experience, and you're starting to go a little bit off track in terms of being less rigorously scientific yourself—in other words, if you're getting kind of subjective—you can draw from BA's example, as he is very consistent in terms of modeling scientific rigor. Which means that he's also willing to recognize shortcomings and mistakes, including his own. You know, nobody goes through life without making mistakes. But he will not just do things on the basis of what's popular, he won't bend himself to popular or fashionable inclinations, or things like that. He keeps coming back to: Where's the evidence, let's look at the process, let's look at what things are rooted in, what would it actually take, and what would it look like, to transform this phenomenon, and so on.

Again, this modeling of consistent, rigorous scientific method and approach is something that I feel I've benefitted from a great deal. It has affected my work a great deal, and my contributions, to the extent that I've made contributions around things like the woman question, or the question of promoting scientific understanding of biological evolution, or popularizing science more generally. As I've said many times, knowing what's real, and why it matters, is extremely important for people broadly in society—to have some understanding of the material world and of where life comes from, where people come from, how all life has evolved. Of course, knowing about all this is very cool in its own right, just in terms of being able to understand life, but it's also very important for societal reasons for people to be very deeply clear that the origins of people, the origins of all life, has nothing to do with supernatural beings. There are natural processes, natural material processes that have taken place—and, again, I think all this is explained pretty clearly in the *Evolution* book [*The Science of Evolution and the Myth of Creationism: Knowing What's Real and Why It Matters*]. But my point here is to say that BA has had an influence, directly or indirectly, in all the dimensions of my work, above all through the modeling of method and approach, and the emphasis on what it is people most need to know about and understand.

An Explorer, a Critical Thinker, a Follower of BA: Understanding the World, and Changing It for the Better, in the Interests of Humanity

Question: I thought a good note to end on would be: What does BA's leadership and new synthesis of communism have to do with how you understand and approach the world?

AS: [Laughs] People sometimes inquire about what kind of people will work with BA or follow his leadership. And I guess that's part of what your question is trying to get at. Well, I would say, just look around. I think you'll find an impressive and diverse mix of creative people of conscience with many different backgrounds, skills, and personalities. Speaking for myself, I guess I'd say that I'll always be a critical thinker. I just don't know any other way to be! [laughs] I'm sure I'll always be curious about just about everything, both in the natural world and in human society. I am both challenged, and sustained, by the diversity and complexity of the natural world and the social world. I think I am, at heart, an explorer. Exploring the unknown, discovering what has not previously been understood, breaking new ground: In my own view, this is a lot of what makes life worth living.

But I also don't want to *just* understand the world. I want to help *change it*, for the better and in the interests of all of humanity. And that's where BA's new synthesis of communism comes in for me. Because thanks to BA's new synthesis of communism, and especially as it is concentrated in his application of scientific methods and approaches, I feel that I have gained, over the years, a much deeper appreciation, not only of the great complexities of the overall process of revolutionary transformation, but also of the very real *possibilities* for such transformation. How you could actually do it. How you could actually win. How you could actually bring into being a new society that would be worth living in.

If it weren't for the new synthesis of communism, I might have gotten discouraged. In my own work on the woman question, in my work on popularizing the science of evolution, and in many other areas where I have tried to make some contributions, I have repeatedly drawn great insights from the new synthesis epistemologically and methodologically, and I have tried to apply this in my work, to good effect, I think. In all of my life's work, I think it's clear that I am very committed to spreading basic scientific understanding and methods among the people as broadly as possible, helping many, including from the most oppressed and the least formally educated, to actually enter into and participate in the scientific process in their own right. And I am also committed to bringing to bear all my training and life experiences to bringing a more consistently rigorous scientific approach into every nook and cranny of the movement for revolution and to forging the pathways that go towards a new society, a new socialist transition towards communism. And BA's new synthesis of communism, and the whole method and approach that most clearly characterizes and concentrates it, has inspired and provoked and challenged my work in many positive ways over the years, and in many dimensions.

Again, more than anything else, it is **the method and approach** concentrated in the new synthesis, and in particular its epistemological dimensions: its rigorous pursuit of the patterns that reveal material reality as it really is, regardless of how unexpected and how uncomfortable those discoveries might be; and its scientific grasp that it is always the contradictions that exist within a thing or process that provide the material basis for change; and that therefore you will find that the **material basis** for the radical, revolutionary transformation of society and the world resides primarily **right within the handful of the key underlying contradictions**, the ones that constitute the core underpinnings and defining characteristics of the prevailing system, which today is the system of capitalism-imperialism that currently dominates the world. All this has not only provided the framework within which I feel one can "ask the right questions," increasingly, but also pursue those questions to their resolution. It has, in a very real sense, provided me personal sustenance and air to breathe. And I feel that it has enabled me to make at least some significant contributions to the overall process of scientific discovery and transformation in various spheres. Not just for my own enlightenment, or because of my own curiosity, although it does assist in this as well [laughs], but also to help advance the process of radical transformation of society that is needed so urgently and by so many. BA's new synthesis of communism has challenged me in positive ways, and enabled me to make contributions that I would not otherwise have been able to make. And, speaking not only for myself, but for many others who have been inspired in their own work and in their own contributions by BA's new synthesis, that once again is a sign, an indication, of what I think of as really good scientific leadership.

***

TABLE OF CONTENTS